South East Essex College
of Arts & Technology

www.ce.strath.ac.uk/nccc

David Langford and Branka Dimitrijević
with
Steve Arnett
Mahtab Farschi
Norman Gilkinson
David Wilkin

Published by Thomas Telford Publishing, Thomas Telford Ltd, 1 Heron Quay, London E14 4JD.
URL: http://www.thomastelford.com

Distributors for Thomas Telford books are
USA: ASCE Press, 1801 Alexander Bell Drive, Reston, VA 20191-4400, USA
Japan: Maruzen Co. Ltd, Book Department, 3–10 Nihonbashi 2-chome, Chuo-ku, Tokyo 103
Australia: DA Books and Journals, 648 Whitehorse Road, Mitcham 3132, Victoria

First published 2002

A catalogue record for this book is available from the British Library

ISBN: 0 7277 3148 3

© D. Langford and B. Dimitrijević 2002

All rights, including translation, reserved. Except as permitted by the Copyright, Designs and Patents Act 1988, no part of this publication may be reproduced, stored in a retrieval system or transmitted in any form or by any means, electronic, mechanical, photocopying or otherwise, without the prior written permission of the Publishing Director, Thomas Telford Publishing, Thomas Telford Ltd, 1 Heron Quay, London E14 4JD.

This book is published on the understanding that the authors are solely responsible for the statements made and opinions expressed in it and that its publication does not necessarily imply that such statements and/or opinions are or reflect the views or opinions of the publishers. While every effort has been made to ensure that the statements made and the opinions expressed in this publication provide a safe and accurate guide, no liability or responsibility can be accepted in this respect by the authors or publishers.

Typeset by the authors
Printed and bound in Great Britain by Bell & Bain Ltd, Glasgow

Srdjan

This monograph is dedicated to you. I miss you, my friend. Maybe, another brandy, another time.

                                                Dave

To Srdjan, my beloved husband.
Your love and support are built into this work.

                                                Branka

Acknowledgements

The Network of Construction Creativity Clubs (NCCC) was initiated and supported by members of the Association of Researchers in Construction Management (ARCOM). The Engineering and Physical Sciences Research Council (EPSRC) provided a one-year grant to support the network.

The Network's successful activity would not have been possible without the dedicated and effective work of the NCCC regional co-ordinators and the ARCOM members at the universities which formed the Network. We are also indebted to all presenters who showcased their innovations at the NCCC events.

At the time of press the ARCOM members steering the network are listed below:

Professor SA Akintoye, Glasgow Caledonian University
Dr D Boyd, University of Central England, Birmingham
Dr ARJ Dainty, Loughborough University
Dr F Edum-Fotwe, Loughborough University
Dr CO Egbu, Glasgow Caledonian University
Mr CJ Fortune, Heriot-Watt University
Mr DJ Greenwood, Northumbria University
Dr WP Hughes, University of Reading
Dr D Hughill, University of Manchester Institute of Science and Technology
Dr F Khosrowshahi, University of Central England
Professor D Langford, Strathclyde University
Professor L Ruddock, Salford University
Dr P Stephenson, Sheffield Hallam University
Dr M Vian, Wolverhampton University

With great dedication and enthusiasm, Ms Marianne Halftorty, University of Strathclyde, helped us to prepare this monograph for publishing.

Contents

Acknowledgements   iv

1	**Executive Summary**	1
2	**Background**	2
3	**Network objectives and results**	4
4	**Innovative companies and organisations**	6
5	**Trends and perspectives**	17

References   23

Appendix – Innovations presented at NCCC events   25

1 Executive Summary

This publication provides a background and profile of innovations in the UK construction industry through an analysis of the innovations presented within the Network of Construction Creativity Clubs (NCCC). The NCCC was a one-year project funded by the Engineering and Physical Sciences Research Council (EPSRC) for the period January-December 2000. The Network was initiated by members of the Association of Researchers in Construction Management (ARCOM). Its aim was to provide a mechanism to exchange knowledge and expertise on innovations between the construction industry, small and medium sized enterprises (SMEs), and the academic community.

Through the network of four regional clubs, which comprised 12 universities across the United Kingdom, the NCCC offered opportunities for informal communication between innovators in each region. In the period January-December 2000 the NCCC organised 23 events. The events provided an opportunity for networking, which was also supported through the NCCC web site by publishing the profiles and contact addresses of the companies/organisations which presented their innovations. The NCCC has thus provided not only a forum for direct knowledge transfer on innovations between the construction industry SMEs and academia, but it provided information on the conditions and mechanisms which support the innovations. This information is an important feedback mechanism towards better understanding of innovation processes in the construction industry.

Analysis of innovations, presented by companies and organisations at NCCC events, provides a sample of the current innovative state of the UK construction industry. The analysis of collected data shows that significant efforts are being made in the construction industry, academia, professional organisations and through government initiatives to stimulate and achieve improvements. The sample shows that innovations are taking place in all construction related areas, but especially in environmental impact management, contracting and partnering, procurement, and application of IT. The innovations are related not only to the processes, products and practices of the industry, but to other areas which have an influence on the construction industry (e.g. insurance, marketing).

The interest of the industry and academia to showcase their innovative work, which was notably growing with each NCCC event, shows that all those involved in the development of the UK construction industry and related areas are making significant efforts to contribute to improvements and to promote their achievements. In the period January - December 2000 the NCCC achieved the following:

- organised 23 events
- attracted 80 presentations on innovations
- attracted over 600 attendees at the events
- published over 50 presentations on the web site
- produced one newsletter.

2 Background

In the last three decades governments in developed economies have been placing greater emphasis on measures to support small and medium size enterprises (SMEs), which are a potent vehicle for the creation of new jobs, for regional economic regeneration and for enhancing national rates of technological innovation (Rothwell, 1986). Governments use a range of policy measures to support and promote innovations in the construction industry. The UK stimulates the rate and direction of innovation in the construction industry by deploying a range of instruments which aim to support:

- R&D - Foresight, EPSRC's Response Mode Research and managed programmes
- advanced practices and experimentation – Partners in Innovation, EPSRC's IGDS courses, Movement for Innovation
- performance and quality improvement – BRE Framework agreement
- taking up systems and procedures – Construction Best Practice Programme, Government Procurement Initiatives, Regulatory system (Winch, 2001).

Dissemination of knowledge on innovative practices in the UK construction industry should take place as a phase in the implementation of wider government policy in this sector. The UK's Foresight programme was first announced in the 1993 White paper "Realising our Potential". Its aim is to identify opportunities in markets and technologies which will enhance the nation's prosperity and quality of life (DTI, 1997). Development of learning networks and creating a culture of innovation, were two of the panel recommendations aimed to sharpen the construction industry ability to respond to changing market conditions and improve its competitiveness (DTI, 1997).

Many innovations in the UK construction industry are developing in response to the latest government initiatives based on Sir John Egan's report "Rethinking Construction" (Egan, 1998), which gave recommendations for performance improvements. The Construction Industry Task Force also concluded that the major clients of the construction industry must provide leadership by implementing projects which will demonstrate the proposed approach. Thus, in November 1999 it initiated the movement for change and radical improvement in the process of construction, named Movement for Innovation (M^4I). The M^4I developed a programme of Demonstration Projects which in its first year attracted 84 projects nominated by the industry (Huntigton, 1999). The M^4I aims to be the means of sustaining improvement and sharing learning by publicising the information on the demonstration projects in professional journals (e.g. New Civil Engineer Supplement, November 1999), at the M^4I conferences and on its web site.

The Construction Industry Research and Information Association (CIRIA) has produced a study which aims to identify the factors necessary for the implementation of learning networks (Holti and Whittle, 1998). The study identifies two main types of learning networks, broker (i.e. a network which represents its members collectively to customers, suppliers and other stakeholders) and thematic (i.e. a network which brings its members together so that they can pursue some common agenda and learn directly from each other) (Holti and Whittle, 1998). The Network of Construction Creativity Clubs (NCCC) was created as a thematic network which brings together academic and construction industry practitioners, and promotes a culture of innovation. A pilot Glasgow Construction Creativity Club (CCC), organised over the course of a 12-month period during 1996/97, showcased local SMEs and University of Strathclyde innovations. Events were held which demonstrated Futures Research Modelling for the Construction Industry; the application of Virtual Reality to construction planning; 3D modelling systems; the quest for creative ideas; the paperless site; and innovative ways of housebuilding. The success of the pilot led Strathclyde University to investigate mechanisms to develop the CCC concept further.

The challenge was to create a culture of innovation by developing a network of CCCs, to facilitate interaction between innovators, users of innovation and scholars. At a meta level the challenge is to create a cultural change. This is easier said than done but there are increasing signals that the benefits of cooperative working are being recognised. Partnering, innovative designs, greater use of separate organisations coming together for design and build projects, the greater use of tool box meetings and induction programmes for sub contractors, are all illustrations of this changing climate. Indeed, PFI programmes, with the necessity of bringing together financiers, designers, contractors, facility managers etc, encourage this new climate of working. Measuring the extent to which construction sector SMEs are responding to this climate of change relies more on anecdotal evidence, rather than a structured assessment. The corresponding Network challenge therefore includes exploration and evaluation of the research and innovation process within SMEs. For the many SMEs, which populate the sector, formal 'academic' research has little to offer unless it is linked to a project or process that offers immediate or short-term paybacks. Formal 'academic' research is also often presented in places and formats unattractive to practitioners and couched in unfamiliar technical terms. Innovation is different. It is seen to be financially and emotionally owned by companies.

The Construction Industry Council's document "Profit from Innovation" points out that more innovative businesses tend to be more profitable (CIC, 1993). The Research Council's endeavours in the IMI programme and the DETR's 'Partners in Innovation' and LINK programmes, have further fostered this view. If innovation offers value and profitability then it is what companies should do. The corresponding Network challenge was to research, evaluate and profile the process and practice of innovative behaviour within construction sector SMEs; and disseminate examples of best practice to less innovative firms.

The outputs are usable in the short, medium and long term. Initial transmission of innovation to Club participants was short term, and perhaps an individualised experience. In the medium term, the electronic and traditional documentation is a record of innovative behaviour and best practice case studies to guide the corporate behaviour of the firms making up the industry. In the long term the programme contributes to changing the culture of the industry.

3 Network objectives and results

Figure 1. Location of the universities which participated in the NCCC

Four regionally based CCCs were established to serve as a focal point for construction sector innovation in each location (Figure 1):

- The Southern CCC was led by: The Department of Construction Management at the University of Reading; and the Department of Construction Management at the South Bank University.
- The Midlands CCC was led by: The School of Architecture at the University of Central England; The School of the Built Environment at Coventry University; The School of Engineering and the Built Environment at the University of Wolverhampton.
- The Northern CCC was led by: The School of Construction at Sheffield Hallam University; The School of the Built Environment at Liverpool John Moores University; Department of Construction Management, Leeds Metropolitan University; and The Department of Surveying at the University of Salford.
- The North East and Scottish CCC was led by: The Department of Building and Surveying at Glasgow Caledonian University; The Department of the Built Environment at the University of Northumbria; and The Department of Civil Engineering at Strathclyde University.

The regional CCC's organised up to 6 meetings over a 12-month period. By the end of December 2000 the clubs had organised 23 events with the participation of 80 presenters. A standard template was used to collate information. Each CCC encouraged the participation of SMEs and identified and put forward examples of innovative practice from their regions/areas of expertise. The principal investigator saught to co-ordinate the overall programme so that a range of innovations are showcased from different specialist sectors (e.g. - housebuilding, civil engineering etc). Innovations also covered different technology levels, and the various stages in the construction process, including research, manufacture, design, assembly.

Each NCCC event usually had a main theme which encapsulated the common link in the presentations. The selection of themes offered the possibility of including topics which are dominant in the construction industry today. Questions and discussions between the presenters and the professionals from the SMEs participating at the event followed all presentations. The events have been an opportunity for networking, and this has been further supported through the NCCC web site by publishing contact addresses of the companies and organisations who present their innovations.

The individuals drawn to the NCCC meetings were typically enthusiastic professionals, from all areas of the construction sector. Following on from this, the structure of meetings and events was informal so that connections were made, innovations and ideas shared, developed and improved upon. The benefits of these events include:

- Providing fora for the discussion of innovations in business (e.g. a typical event showcases the innovation; evaluates the reasons for its success; and considers its relevance to participants and a wider construction sector application).
- Providing opportunities for networking to take place e.g. to facilitate the flow of ideas from one area of business to another.
- Providing a framework for cross fertilisation of ideas.
- To SMEs: Exposure to different innovations and being part of a growing culture of innovation in the industry and an opportunity to 'partner' innovations together.
- To individuals: Exposure to innovators, seeing how they work, what motivates them etc. This benefit is expected to be transdisciplinary in that the CCC's are open to a wide range of disciplines connected to the built environment.
- To academic partners: Exposure to innovative companies and individuals who welcome ideas drawn from research. The academic partners also benefit by being exposed to research and innovation mechanisms being used in other higher education institutions.

Audience levels were approximately 20 SME representatives per meeting. The network as a whole attracted over 500 business participants, and exposed them to innovation and shared ideas, that may improve the performance of their companies. A newsletter has been produced to profile events and innovations. The project website has been used to house the presentations of innovations, to enable electronic dissemination of outputs and facilitate further communication and networking between participants and the construction sector.

4 Innovative companies and organisations

This section provides analysis of information on innovators and innovations, collected during the NCCC programme of events. The data and information were collected using the following methods:
- at the NCCC events during the innovation presentations
- at regular NCCC steering group meetings (exchange of experience)
- through written feedback reports from NCCC co-ordinators (innovation pro-formas / innovation questionnaires).

The NCCC web site offers a complete listing of 80 innovations, companies and academics who presented them, and a summary of the presentations which have been made available so far. The completed innovation pro-formas were obtained for 35 innovations.

Key points of the analysed information can be summarised as follows:
- Presenters at the NCCC events were from industry (61%), academia (31%), professional organisations (3%) and government initiatives (5%).
- The highest number of presentations were about environmental impact management (8 presentations), procurement (7), and contracting and partnering (7).
- Among those who provided the information on financial turnover, 11% had a turnover of less than £150,000; 3% between £150,000 and £500,000; 11% between £500,000 and £1m, 36% over 1 M. Further 28% belong to the public sector, and 11% withheld this information.
- Regarding the number of employees, 22% of organisations did not provide this information, 43% have less than 250 employees, 6% less than 1000, and 29% more than 1000 employees.
- Different types of partnering in developing innovations feature in 17 of the 35 companies who provided this information.
- Lead companies provided 100% of funding in 17 companies, while 18 of them obtained funding from additional sources.
- The needs of client were the most important origin and driver of innovation. Thus, the innovators perceive that the clients will be the main beneficiaries of innovations.
- Among the companies who provided the information, 68% declared that their innovations are not subject to patent.
- While 76% of companies encountered a range of difficulties in the conception, development and implementation of their innovations, 26% did not have any difficulties.
- With regard to the dissemination of innovations, 57% of companies have a dissemination structure in place. The main reason for not having a dissemination structure was identified as 'the innovation being specific to the business' (23%).

- The largest percentage of innovators (27%) consider their innovations to be transferable to a wider construction industry, clients (21`%), other sectors (19%), subcontractors (15%), and other groups (5%).

An analysis of the type of companies and institutions who contributed their presentations shows that a wide range of organisations have shown an interest in promoting and disseminating innovations in the construction industry (Figure 2.)

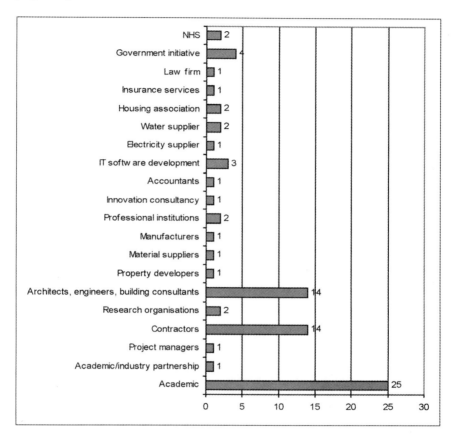

Figure 2. Business activity of innovators

NCCC events have been a forum for sharing information on innovations between academia (31% of presenters), industry (61%), professional organisations (3%) and government initiatives (5%) (Figure 3.). Architects, engineers and consultants (18%), and contractors (18%) formed the largest group of presenters from the industry. However, many other sectors of the construction industry and related professional areas were interested in presenting their innovations as well, e.g. developers, material suppliers, public investors (NHS), building services, manufacturers, software developers, accountants, insurance, research institutions. EU and UK professional organisations, and UK government initiative for support and dissemination of innovations (M^4I) have presented their programmes which support innovations.

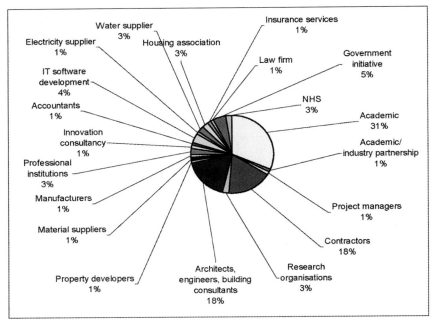

Figure 3. Business activity of innovators

An analysis of the types of innovations presented at the NCCC events (Figure 4.) shows that improvements are taking place not only in the construction industry, but in the areas related to this sector (e.g. insurance, marketing) as well.

The highest number of presentations was about environmental impact management (8 presentations), procurement (7), and contracting and partnering (7). Sustainable technologies, safety and risk management, construction process, and financial management were topics of four presentations in each area. Three presentations covered each of the following topics: training, teamworking, energy efficient design, innovation in SMEs, general case studies of innovations, liability legislation and insurance, government initiatives for improvements to the construction industry, and IT in information strategy plan. Product standardisation, product innovation, innovative housing, and virtual reality were topics of two presentations in each area. One presentation covered each of the following topics: feedback on innovations, social issues, marketing, developing opportunities, design and productivity, asset management, performance measurement, quality control, virtual organisations, and IT in estimating.

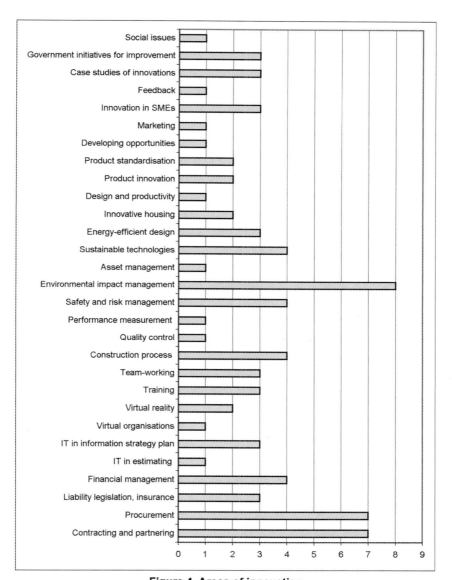

Figure 4. Areas of innovation

Competitive advantage is being sought in addressing issues of sustainable construction through building design, construction process and building management. The efforts in improving the efficiency of the industry focus on new types of procurement, contracting and partnering, the application of IT, and training with regard to new technologies and team-working. The variety of other presented topics illustrates the development of innovative approaches and solutions in all areas of the construction industry and related fields.

The complete information has been collected on 35 of 80 presentations. The analysis shows that the ownership of the companies/organisations presenting their innovations was as follows: 48% are privately owned, 29% come from public sector (the higher education and NHS), 11% are part of a corporation, 6% are a Plc, and 6% could not be classified in these categories (Figure 5).

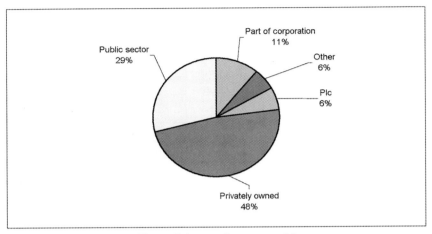

Figure 5. Innovation ownership

The figures on financial turnover of the companies/organisations presenting their innovations could not be completely provided because 11% of them withheld the information as confidential, and a further 28% of organisations belong to the public sector (universities, NHS) (Figure 6). Among those who provided the information 11% had a turnover of less than £150,000; 3% between £150,000 and £500,000; 11% between £500,000 and £1m, 36% over 1 M.

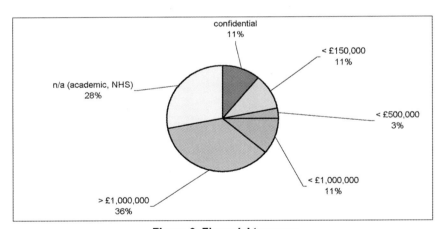

Figure 6. Financial turnover

Among the 35 companies, 19% have less than 10 employees, 8% have between 10 and 50 employees, 8% between 50 and 100 employees, 8% between 100 and 250 employees, 6% between 250 and 1,000 employees, 29 % over 1,000 employees. 22% did not provided this information (Figure 7). The readiness of small and medium size companies to develop and promote their innovations indicates the importance placed on innovative approaches in order to achieve competitive advantage.

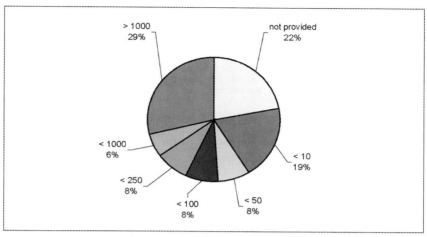

Figure 7. Number of employees

Partnering in developing innovations features in 17 of the 35 companies who provided this information. The number of partners varies from 1 to 14, and in one case there are over 150 partners. The latter example is of an IT consultancy which collaborates with the regional businesses in finding innovative ways to use e-commerce to expand. In 37% of companies/organisations the innovation was undertaken as a sole venture; 31% formed academic/industry partnerships; 9% joined in partnering ventures; 9% opted for commercial joint ventures; and 14% formed other partnership relationships (e.g. non-commercial public-private partnership), or partnering could not be formally defined (e.g. government initiatives) (Figure 8).

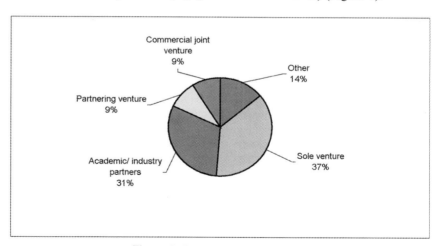

Figure 8. Partnership mechanism

Lead companies provided 100% of funding in 17 out of 35 companies/organisations. Other sources of funding are presented in Figure 9.

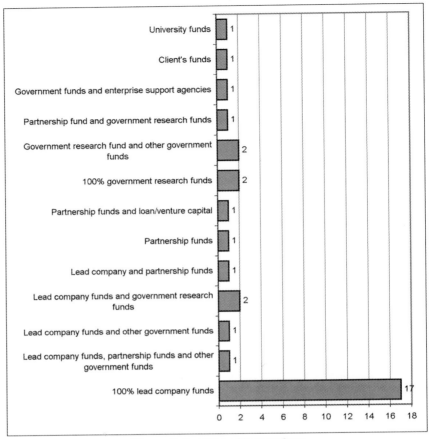

Figure 9. Funding of innovations

The innovators identified the origin of innovations as one or more of the following: needs of the site; needs of business; needs of industry; needs of client; needs of market; research initiative; competition (e.g. for sustainable housing); and personal beliefs/ideas on business operation. Although the innovators were asked to put the origins of innovations in order of importance (e.g. 1 for most important, 2 to second in importance, etc.), some of them have assessed that the origin of innovation can be equally attributed to all or a few of them as the most important cause (Figure 10). The needs of client were the most important origin of innovation, followed by the needs of the industry, needs of business, research initiative, needs of market, needs of site, or a combination of several of these needs. Design competition (put forward by the client) and "somebody's idea" have also been mentioned as the most important origin of innovation.

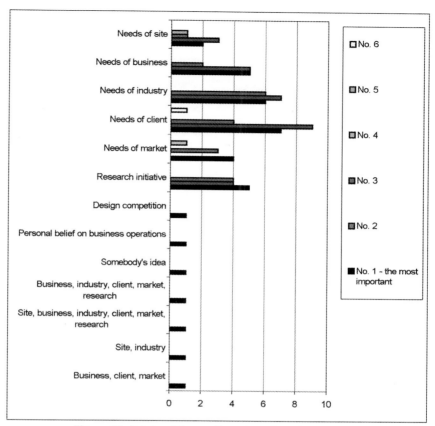

Figure 10. Origin of innovations in order of importance

The main drivers of innovations (Figure 11) were identified as follows: client requirements, followed by enhancing competitiveness, internal efficiency, embracing IT, university research, sustainability of construction industry, government policy, combination of these factors, or personal interest and research.

Innovators perceive that the following groups will benefit from their innovations (in order of importance): clients, the wider industry, the innovative company, partnership organisations, subcontractors, the company's workforce, all stakeholders, local community, and users (e.g. tenants) (Figure 12).

Among 35 companies who provided the information, 68% declared that their innovations are not subject to patent. When asked if they encountered any difficulties in the conception, development and implementation of the innovation, 26% of innovators said that there were no difficulties, and others listed a wide range of difficulties which are presented in Figure 13.

14 *Innovative companies and organisations*

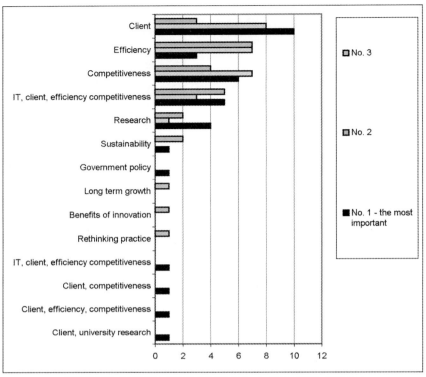

Figure 11. Drivers of innovations in order of importance

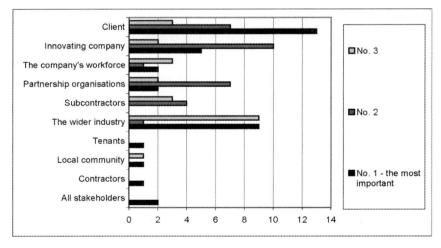

Figure 12. Beneficiaries of innovations in order of importance

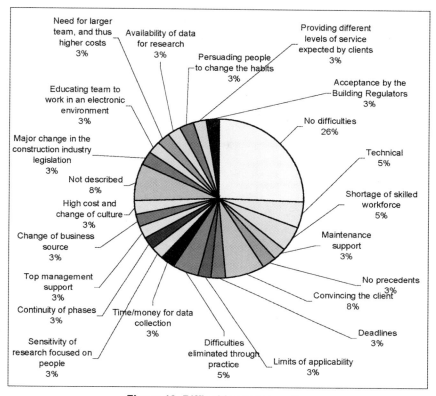

Figure 13. Difficulties in innovating

With regard to the dissemination of innovations, 20 companies among 35 who provided the information have a dissemination structure in place. Those who disseminate the innovations distribute this information within the lead company (21%), within the partnership (16%), to subcontractors (8%), to clients (18%), to wider industry (21%) and to other groups (16%) (Figure 14).

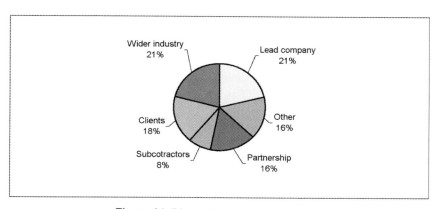

Figure 14. Dissemination of innovations

Companies who do not have a dissemination structure in place (15) identified the following reasons: lack of funding (6%), no active policy (11%), because the innovation is not transferable (12%), because the innovation is specific to the business (23%), because normal sales channels are being used (12%), because industry ignores the experience of other countries (6%), because it is not appropriate (6%), and because the innovation is perceived as a competitive advantage (12%) (Figure 15).

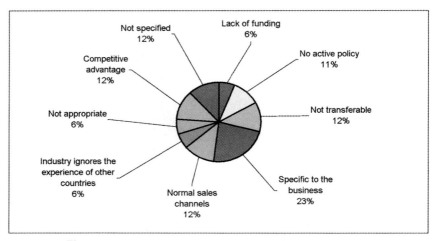

Figure 15. Reasons for the lack of dissemination structure

Innovators consider their innovations to be transferable to clients (21%), partners (11%), subcontractors (15%), wider construction industry (27%), other sectors (19%) and other groups (5%) (Figure 16).

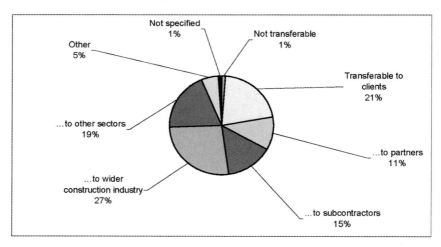

Figure 16. Transferability of innovations

5   Trends and perspectives

The innovations presented in this book represent a cross section of those that have been presented during the life of the National Construction Creativity Club project. For the purposes of the project an innovation was defined as taking place when practices are so new that the set pattern of accepted processes or product is developed or replaced. The introduction of this new process or product requires deliberate action and control.

In this project the team sought to explore the process of innovation through the investigation of innovations themselves. By looking at the nuts and bolts of what a innovation seeks to do we hope to find out more of the spurs which convert ideas into innovative products, services, or processes for the construction industry. At the heart of understanding the innovation process is the linkage between research and innovation.

Research & Innovation

It is well recognised that formal research spend as a percentage of turnover in construction is low and has been estimated as being something like 0.001% of the output of the industry. White goods companies would typically spend 20% of their turnover on research. Does this suggest that as construction does not, then it is a backward industry? Groak (1992) has argued the reverse. The construction industry is innovative but not driven by formal research programmes. The innovation which takes place is informal, unrecorded and bespoke to one project. It takes place in the pub, over the drawing board or computer and is stimulated by the need for an innovative response to an immediate problem. However, programmes which aim to encourage innovation in the construction sector are also being developed such as the French system promoted by a specific institution: Plan Construction et Architecture (PCA) (Campagnac, 1998). PCA encourages innovation by experimentation by mobilizing all the representatives of the construction sector (clients, professionals, contractors, etc.) to conceive and work together on innovation. The results of these experimentations are evaluated by independent researchers.

In relation to strategies for developing innovations in other sectors, which can be applied to the construction industry, Dulaimi and Kumaraswamy (2000) point out that the process of using R&D to deliver successful innovations in manufacturing industries is similar to the development process in construction from concept, to detailed design and construction. In order to ensure a successful product to any potential customer, the manufacturing industry seems to pull the different parties and disciplines together more strongly. Dulaimi and Kumaraswamy (2000) argue that the integrative pull of innovation process could be beneficial in construction procurement systems, and related operational, educational-training and technological systems. In depth case studies of small construction companies are being undertaken to understand the role and significance of innovations for them, and especially the potentially adverse implications of shot-term innovation to the detriment of more long-term, strategic innovation (Sexton et al, 2001).

This project has attempted to capture some of these innovations and so uncover the innovative process. Clearly this approach is matched by Governments' desire to 'modernise' the construction industry and has dirigistically sought to capture innovations being used in construction through its Movement for Innovation and the associated demonstration projects. Instruments of state have been bent to serve this modernisation process; the DETR sets the agenda, announcing what it wants to fund and the research councils policies serve this agenda. It can be seen that there is now concerted efforts to integrate research and innovation where research is harnessed to practical rather than critical outcomes. In order for this transfer from research to commercialised innovation to be enabled we have seen is the pulling together of government, industry and the universities to form a powerful axis driving the research agenda for the industry Green (1999) sees the model as being supportive of clients efforts to get 'value for money' in their construction projects. Fig 17. illustrates this confluence of industry, universities and government and the relationships that the industry sees as needing to be in place to promote innovation.

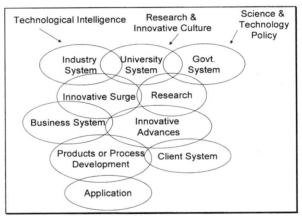

Figure 17. Innovation environment

The model highlights several symbiotic relationships
- Advances in innovation come through the research and innovation surge systems
- Improved performance in products and processes come from interfaces with the innovation advances which emerge from the business system and are encouraged or enabled by the industry's clients.
- Decisions about the use of innovation lie in the business system but evaluations of the innovation will be taken by clients who will monitor the innovation in reducing costs and time and improving quality or whatever strategic objectives they have.

It is interesting to note the spread of innovations being presented through the NCCC. It is stressed that whilst the innovations showcased in this report were selected by the regional co-ordinator the basis of selection was somewhat random and largely depended upon the willingness of speakers to present their work. Consequently the sample may be said to random(ish). In this context the groupings of innovation reflect the interests of each of the regional clubs of the NCCC.

The Environmental Context of the Innovations

The innovations showcased in this report were all presented in 2000. It may be argued that if the economic conditions are right then innovations are more likely to take root. The shake up of the industry, driven by government has fostered considerable change and the government and industry encouragement has spawned organisations such as the Movement for Innovation and the Construction Best Practice Programme which have provided a backcloth to encourage innovation. Moreover more firms sought to gain competitive advantage from the use of innovative ideas. It is noticeable that the needs of the client was the dominant driving force for innovation.

The period leading up to the presentation of the innovation could be found in Lansley's (1994) ideas about the nature of the environment and managerial behaviour. Lansley saw the environment evolve from an operational environment (1960 - early 70's) which was characterised by stable demand and high levels of public expenditure on construction. This matured into a strategic environment (1970 - 80's) which displayed features of turbulent change through to the competitive environment of the early 80's -90's in which demand was unpredictable. The period leading up to the NCCC project may be said to fit the strategic environment with its restless search for changes in procurement, value propositions for clients and contractors, experimentation with new materials and building technologies in a setting where sustainable construction products and methods are demanded.

In all the environment is one in which combines features of dynamic change within the general framework of features of economic expansion. The changes were brought about by Latham and Egan, coupled with the DETR encouragement in promoting a modernised construction industry. Hence the construction environment could be seen as somewhat discontinuous and unpredictable in terms of changes in construction practice if not the economic setting. In such an environment firms will seek out flexibility in operations to cope with novel changes in construction practice.

Creativity will be paramount and open systems which seek to use soft boundaries between the parties involved in a construction project will be used to achieve strategic effectiveness. This condition is reflected in the source of the innovations recorded. Some 21% of the innovation were championed by firms only tangentially connected to the business of construction. What is also evident is that Universities have a strong position in linking innovation to practice and 32% of the innovations can be seen to emanate from academic institutions. Using a model developed by Pries and Janszen (1995) we can seen the flow of innovations that were presented in this project (Fig. 18). The percentage of innovations from each source is given in each box.

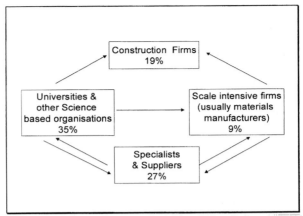

Figure 18. The flows of innovation After Pries & Janszen (1995)

The external environment is also shaping what is being innovated. This reinforces the view that when external conditions are right the more innovation is possible. If a subject is under close public discussion it is likely that innovative activity will take place. The model shows the flow of innovations. Specialists suppliers have provided 27% of the sample and are likely to have worked up ideas with Universities and other science based organisations and firms which depend upon large scale operations, (usually bulk materials suppliers). The Universities are likely to provide the science base for these material suppliers as well as innovations for the contractors to develop. (any number of expert systems and software innovation have a foundation in University construction departments). Equally the materials manufacturers find themselves providing product innovations used by construction firms.

It is this multi-organisation approach which sponsors much innovation. Kangare and Miyataki (1997) see the formation of cross industry strategic alliances as the most powerful combination to promote innovation. Fusions of technology drawn from various disciplines also encourage innovation. In our sample there was evidence of strong cross company collaboration to promote innovation but more evidence of innovation being enabled by mixtures of private and public funding. An inspection of the areas of innovation we see that the issues dominating the construction industry agenda in the 1990's are well represented in the NCCC sample. Such that environmental management leads the list, closely followed by procurement and partnering. Again it is stressed that the innovations presented in this project were not representative of all innovation taking place it is indicative that the nature of the external environment shapes the kinds of organisations involved in innovation and to what purpose this creativity is put.

Impediments to Innovation

Gerwick (1990) identified several barriers to innovation. Principal amongst them was the short term return expected by innovation. In his paper he cites an Economics Professor (D Tesse) who claims that the social rate of return of a fully developed innovation is 4-10 times greater than the direct rate of return.

Whilst cautioning against evaluating innovation by standard economic tests Gerwick recognises that innovation does take time to develop and trial innovations and this may be costly. In our sample the impediments to the innovation are extremely diverse with 24 factors being cited as 'difficulties'. Of these 'difficulties' 19 were unique in that the difficulty was only faced by one innovator. This data gives us an insight into particularities of innovation.

On the issue of the costs of innovation the sample we have indicates that innovation are, in the main, prepared to fund their innovations. The largest number of innovations in the sample, by a long way, (17) are funded by the companies themselves; only 9 innovations have been prompted by wholly or partially funded by government schemes.

Surprisingly, given the advice from Japan that strategic alliances are the central model for promoting innovation, only 4 innovations are based upon partnerships. Although when an innovation is ready for dissemination some 12 firms use partnerships to activate the innovation. This tells us something of the process of innovation. Having noted this only 57% of the innovations had a defined innovation structure in place and it is likely to individualistically championed and tested in its early stages then spun out for wider trials across organisations who have a stake in the outcome of an innovation. The Japanese model of innovation by a 'committee' of partners does not seem well suited to the practice of innovation in the UK. Individualistic and competitive motives may pervade the innovative process in the UK whilst more co-operative values may inform the Japanese culture.

The issue of company size may have a bearing on the promotion of innovation. Japan has an industrial structure which is dominated by large corporations who see benefit in R&D and patenting their discoveries. The industrial structure of the UK construction industry is different whilst smaller companies are more populous than the larger companies the lions share of work goes to the large firms. Looking at the wellspring of the innovations we see that 35% of the inventions come from firms with under 100 employers 14% from the middle group of 101-1000 employers and 29% from 1000+ companies. (22% of the sample did not declare the company size). The issues of size also is played out in the turnover of the firms being represented through the NCCC. Excluding the academic presenters it is noticeable that the over £1,000,000 turnover company represents the largest group of innovators. So we see two innovative groups -large firms and smaller firms.

The large companies are investing in innovation as a vehicle to support their reputation as being innovative and the smaller firms building a reputation through innovation. The medium sized firms are in a sense trapped in the middle, neither having resources nor personnel to innovate. These are of course gross generalisations and what is likely to produce innovation is one of a culture where failed attempts to innovate are tolerated and innovation champions are encouraged and rewarded.

Who are the innovators?
There are three dominant sources of innovation in the sample; construction designers (17%), contractors (17%) and academics (31%). Men presented 48 of the 52 innovations cited in this monograph. Why should this be so? The innovators are strongly influenced by their immediate social and working environment. And the imbalance in the gender of the innovators does not mean

that women are inherently less creative than men but speakers more of the social expectations for women. Due to these conditioning factors they are less likely to engage in "technically inventive activities." These are the very conditions pertaining to the construction industry. Most of the innovators are not drawn from the classical Heath Robinson tradition of garden shed inventors and the sample in this study are invariably bound to some employment framework. Nurturing the creative individuals inside larger firms is a viable stratagem to encourage innovation. Many of the sample were drawn from the research and consultancy practices specialising in bespoke design solutions.

An inspection of the presenters in the sample shows that the individuals are characterised by what Lawson (1980) sees as a 'reservoir' of experience and technical ability. Moreover they appear to be able to synthesis ideas from different sources. In short the innovators were able to be sufficiently agile to use their industry knowledge but express it in different patterns. Given that the innovations are seeking to impress clients (see earlier) the innovators outside of the academic centres are influenced by the market pull theory of innovation - the activity is sparked by some fiduciary or production advantage. Yet the institutional (government, M4I, CBPP, etc) framework for creating innovation has greased the route by which these innovations get to the market.

Conclusion

From our year long experiment it is possible to draw some tentative conclusion. These are:

1. Innovation is seen as increasingly necessary for the creation of competitive advantage especially as clients seen as the major beneficiaries of the innovation.
2. That the innovative culture which is being encouraged is being helped along by strategic partnerships between players in a project or a companies supply chain. If these players can be co-located in say, project offices, then the innovative process is likely to be strengthened.
3. That industry/ academia partnership is a fertile territory for the promotion of innovation.
4. Innovations come from either the small or the large firms. Smaller firms will have innovation through 'design' and are more likely to deliver new or improved component or innovations in design solutions. Larger firms are more likely to innovate either in the organisation of project delivering or in IT applications.
5. The business environment in which innovation is all important. Stability encourages innovation but discontinuous changes enforce the search for creative solutions to old problems.

References

Campagnac, E. (1998). National system of innovation in France: Plan Construction et Architecture. *Building Research and Information*, 26/5, 297-301.

Construction Industry Council (CIC) (1993). *Profit from Innovation*. CIC, London.

Department of Trade and Industry (DTI) (1997). *Winning through Foresight: Action for Construction*.

Dulaimi, M. and Kumaraswamy, M. (2000) Procuring for innovation: the integrating role of innovation in construction procurement. *Proceedings of the 16th Annual ARCOM Conference*, 1/303-312.

Egan, Sir J. (1998). *Rethinking Construction*. www.construction.detr.gov.uk/cis/rethink, 26.9.2000.

Gerwick, B. (1990). Implementing Construction Research. *Journal of Construction Engineering and Management*. 16/ 4. 556-563.

Green, S. (1999). Partnering: The Propaganda of Corporation. *Journal of Construction Procurement*. 5/2, 177-186.

Groak, S. (1992). *The Idea of Building*. E&FN Spon, London.

Holti, R. and Whittle, S. (1998). *Guide to developing effective learning networks in construction*, CIRIA, London.

Huntigton, I. (1999). This is M^{4}I, *New Civil Engineer Supplement*. November, 2-5.

Kangari, R. and Miyataki, Y., (1997). Developing and Managing Innovation Construction Technologies in Japan. *Journal of Construction Engineering and Management*. 123/1, 72-78.

Lawson, B. (1980). *How designers think*. Architectural Press.

Lansley, P. (1994). Analysing Construction Organisations. *Journal of Construction Management and Economics*. 12/4, 300-307.

Pries, F. and Janszen, F. (1995). Innovation in the Construction Industry: The Dominant Role of the Environment. *Journal of Construction Management and Economics*. 13/1, 43-51.

Rothwell, R (1986). The role of small firms in the emergence of new technologies. *In* Freeman, C. *Design, Innovation and Long Cycles in Economic Development*. Frances Printer (Publishers), London.

Sexton, M., Barrett, P., Miozzo, M., Wharton, A. and Leho, E. (2001) *Innovation in small construction firms: is it just a frame of mind?*. Proceedings of the 17th Annual ARCOM Conference, 1/527-536.

Winch, G.M. (2001) Innovation in the British Construction Industry: The Role of Public Policy Instruments, in Manseau, A. and Seaden, G., *Innovation in Construction: An International Review of Public Policies.* Spon, London, 351-371.

Appendix
Innovations presented at NCCC events

Appendix
Innovations presented at NCCC events

	Index
Legal, financial and procurement innovations	
Steven Mason, *Innovative application of downstream alliances*	29
Gordon H. Bateman, *Thames Water's Procurement Strategy*	30
Ian Smith, *The North Tyneside Partnering Agreement*	31
Procurement methods	
Allan Neal, *Recent Innovations in Supply-Chain Management*	32
Robert Francis, *Developing an IT business*	33
Michael Sharpe, *Supporting best e-business practice*	34
Lamine Mahdjoubi and Junly Yang, *A Fuzzy Decision Support System for Materials Routing in Complex Construction Sites*	35
Margaret-Mary Nelson and Marjan Sarshar, *Process improvement in the facilities management*	36
Liability legislation, insurance	
John Goodall, *Is it not time for the British Construction Industry to start benchmarking itself against its more efficient continental European peers?*	37
Victoria Joy, *Contaminated Land: The New Regime*	38
Ray Robinson, *What Can Latent Defects Insurance Do for the Construction Industry?*	39
Financial management	
Henry A Odeyinka and Dr John G Lowe, *The Development of an expert system to manage construction cash flow and associated risks and uncertainties*	40
Farzad Khosrowshahi, *Refinements in Project Financial Management*	41
IT in construction industry	
Mark Shelbourn, *IT Self Assessment Tool*	42
Carl Abott, *How to Develop an Information Strategy Plan*	43
Ghassan Aouad, *Open Systems for Construction*	44
Robert Shiret, *Virtual reality presentations in business: using VR to win more business*	45
Training and team working	
Guillermo Aranda, *Priming Novice Site Operatives Using Navigable Movies*	46
Ian Smith, *Working with multi-cultural, multi- disciplinary team*	47
Construction process, safety and risk management	
Matthew Finnemore, *Structured Process Improvement for*	48

Construction Enterprises (SPICE)
Andrew Fleming, *Process Protocol 2* — 49
Eric Johansen, *The Government KPIs and North Tyneside's Detailed Performance Indicators* — 50
Peter Thompson, *The Evacuation Modelling Software: Simulex* — 51
Iain Cameron, *Construction 'Total Safety Management': A Benchmarking Framework* — 52
Gregory Carter, Simon Smith and Jim Turnbull, *IT Tool for Safety Risk Management* — 53

Environmental impact
Paul Yaneske, *Greencode* — 54
George Pye, *What is Green Electricity?* — 55
John Mullholand, *Energy and Water Management at LMU* — 56
Richard Grey, *Continuous Commissioning* — 57
David Taylor, *A Corporate Environmental Policy* — 58
Branka Dimitrijević, *Durability, Adaptability and Energy Conservation (DAEC) Assessment Tool* — 59
Eric Whale, *Space: A Technology Source* — 60
John Gilbert, *Sustainability in architectural practice* — 61
Douglas Taylor, *New Energy Efficient Housing in Ayr* — 62

Product innovation and standardisation
Hassan Al-Nageim, *New Product Innovation* — 63
Graham Meller, *A Standardisation Trial for the Analysis of Asphalt by the Ignition Method* — 64

Developing opportunities
Graham Woodall, *Seeing Opportunities* — 65
Chris Smith, *Marketing through clients* — 66

Innovation and feedback
John A. Cantwell, *Innovation in Smaller Firms* — 67
Emma Buxbaum, *Raising the competence of SMEs in the Construction Industry* — 68
Colin Pearson, *Feedback for better design and construction* — 69
Mr. Andrew White, *The Sources and Enabling Factors of Innovation* — 70

Note: Appendix includes the summaries of 42 presentations which were provided by the innovators in time for publishing this report. Information about other innovations presented at the NCCC events can be found on the NCCC web site: www.ce.strath.ac.uk/nccc.

Mr Steven Mason, University of Wolverhampton

Innovative application of downstream alliances

Challenge

Recent attempts to address the industry's problems i.e. Latham and Egan have prompted and cajoled construction players to review practices and procedures in their entirety with a mind to undergo a culture change. Partnering and collaborative relationships are one of the cornerstones of this mind change, primarily in order to obtain a win-win situation for the whole team. The research challenge was to investigate the application of alliances within the construction supply chain with particular emphasis on the downstream relationships whether they exist, do they work, what are the benefits and pitfalls.

Objectives and methodology

Objectives were to:
- Examine the concept and history of partnering
- Critically analyse current literature surrounding partnering and alliancing
- Identify supply chain properties
- Examine types of relationships within an alliance
- Establish degrees of interaction for these relationships
- Discuss the drivers and key issues by which the relationship operate
- Discuss legal issues relative to subcontractor contracts
- Establish the benefits and pitfalls of the relationships
- Snapshot of American experience of partnering

Research methodology was via three means: literature review, structured questionnaire, and semi structured interviews

Conclusions

Key learning points from this study have shown that alliancing in the UK construction industry is extremely difficult to achieve and be successful and demonstrated that all parties to an alliance have different aims and objectives. An alliance will stand or fall on the ability of the partners to overcome or cope with behavioural problems along with the issue of how to convert a contractual relationship to one of mutual trust. The opinions of the industry players can serve as a contribution to the overall awareness of how alliances originate and work each component part of the alliance can be a springboard for further research and may lead to a greater understanding of the process steps required in order to set up and maintain a successful alliance.

Future development

- Examination of case studies in particular performance measures
- Selection of a partnering strategy
- Contract selection suitable for underpinning an alliance
- Alliancing process steps
- Balance of trust, money, culture

Contact presenter: Mr Steven Mason
University of Wolverhampton
c/o Mr K.F.Potts, School of Engineering and the Built Environment
Wulfruna Street, Wolverhampton, WV1 1SB
Tel: 01902 321000 ; Fax: 01902 322680
E-mail: steven@pennw.fsnet.co.uk
Web site: www.wlv.ac.uk

Mr Gordon H. Bateman, Thames Water

Thames Water's Procurement Strategy

Challenge
The water industry Regulator has set some tough challenges for the current regulatory period (2000 - 2005), including an initial reduction of 11% in real terms in Thames Water's Utility company's income. The challenge of delivering capital investment to higher standards and for less cost is being met by the development of earlier experiences of partnering and alliancing.

Innovative approach, implementation and benefits

The development of the approach started in 1987 with the need to achieve compliance with the Control of Pollution Act in respect of several major sewage works. Soon after, in the run-up to privatisation, a doubling of capital investment within 12 months prompted the establishment of an "Extended Arm" project team with a contractor and his design subcontractor. A key feature of both elements was a move to target cost contracts based on the Institution of Chemical Engineers Green Book form of contract. The advent of the EC Utilities Procurement Directive on 1 January 1993 led to a major reduction in the number of approved contractors admitted to a new Directive-compliant register. At much the same time new Framework Agreements were put in place with suppliers of key items of plant and materials, facilitating standardisation and reduced prices.

In the period 1995-2000 partnering was developed with two contractors and their designers across a £100M programme of projects on sewage treatment works outside London. Openness and trust were key. The contractors were selected against a range of "soft" issues, as well as ensuring their technical and financial capabilities. Called the "EQUIP" programme, this work achieved M4I Demonstration Project status. Other partnering projects were undertaken during the same period, both "programme" partnering and single project partnering - the latter for major capital works. There was a very significant move from lump sum to target cost contracts over the period.

In April 2000 the water companies in England and Wales entered their third regulatory period, in which the Regulator set more demanding targets in terms of capital efficiencies. This was foreseen, and a new register of alliancing contractors was established in 1999. From that register a process of further pre-qualification and tendering established two key alliances, each of three contractors and their design consultants working with Thames Water. Three will work on treatment works development, and three on networks (water mains, sewers etc) to develop and deliver some 80% of the capital investment managed through the Company's Engineering division over the five year period. Major contracts (> £20M) will be managed separately, though still through an alliancing approach.

Thames Water and its contractors are together developing the alliances, their management, their development of the supply chain, with continuous performance assessment and the Construction Best Practice Programme's Key Performance Indicators to provide some measures. Thames Water is confident, based on past experience, that these new arrangements will meet the challenges of increased outputs for less cost.

Contact presenter: Mr. Gordon H. Bateman
Thames Water, Gainsborough House, Manor Farm Road
Reading, Berkshire RG2 0JN
Tel: 0118 923 7492; Fax: 0118 923 7964
E-mail: gordon.bateman@thameswater.co.uk
Web site: www.thames-water.com

Mr. Ian Smith, University of Northumbria

The North Tyneside Partnering Agreement

Challenge

Apart from some more recent reports which outline the necessary improvements in UK construction industry (e.g. Latham's report 1994, Egan's report 1998), as far back as in 1963 the Banwell Report stated: "We consider that the most urgent problem which confronts the industry is the necessity of thinking and acting as a whole. It has come to regard itself as a series of different parts, roughly consisting of specialist advisers, contractors and suppliers and operatives of various crafts and skills .... these attitudes must change."

Innovative approach and solutions

North Tyneside Schools Partnering Agreement initiated a change in contracting arrangements by developing a Partnering Charter. A Partnering Charter Workshop was held on 14th and 15th of July 2000, with the participation of 37 representatives from 4 organisations. The aim was to identify and discuss issues that would affect the way the partners would work together and reach agreement on a set of words that would form a Partnering Charter. This was achieved by considering working relationships within the following key areas: ethos, philosophy, process and performance.

Implementation

The partners have instigated a programme of workshops to allow all levels within their organisations to broaden understanding of the process and develop Detailed Performance Indicators that will be used by them to measure improvements in performance. The University of Northumbria at Newcastle is assisting the partners establish benchmarks from which to measure improvement. The government initiative Movement for Innovation has established seven working groups that gather and disseminate information on innovative practices. Working Group 5 has focuses on the following issues: demand/supply side of relationship, client role, partnering and team work. Partnering is innovative involving new working relationships and practices and the North Tyneside Schools Partnering Agreement has achieved a Vision 2000 award and 'demonstration project' status with the Movement for Innovation allowing the opportunity for others to learn from the partners experiences.

Benefits

Potential benefits of partnering range from elimination of the "blame culture" to significant reductions in costs and time together with significant increases in quality and health and safety. Partnering can remove the waste of procurement/contractual processes, indecision or revised design and conflicts and disputes.

Future development

The partners are looking to extend the implementation of their agreement by considering the potential for partnering under a PFI regime. The University of Northumbria at Newcastle is undertaking significant research in investigating aspects of the development and performance of the initiative.

Contact presenter: Mr. Ian Smith
University of Northumbria at Newcastle
Ellison Building,
Newcastle upon Tyne, NE1 8ST
Tel: 0191 227 3534; Fax: 0191 227 3167
E-mail: ian.smith@unn.ac.uk
Web site: online.unn.ac.uk

Mr. Allan Neal, AMEC Capital Projects

Recent Innovations in Supply-Chain Management

Challenge

Ministry of Defence says: "Despite numerous reviews of the construction industry criticising the fragmentation, adversarial attitudes, inefficient use of labour, wastage, high cost of construction and functional inefficiencies of buildings, little has changed."

Innovative approach and solutions

AMEC took part in a Construction Supply Network Project with the MoD which developed the Prime Contract procurement model which the MoD is starting to apply across its whole business. AMEC is applying and developing the tools championed in the prime contracting model across its own business. Prime contracting replaces short-term contractually driven single project adversarial inter-company relationships with long term, multiple projects relationships. It is based on trust and co-operation and considers whole life implications. The seven principles of this contracting arrangement are as follows:

- Compete through offering superior value
- Establish long term relations with key specialists
- Make value explicit: design to meet functional requirement for through life cost
- Manage the supply chain through specialist clusters
- Involve the supply chain in design & cost development using target costing, value management and risk management
- Develop continuous improvement within the supply chain
- Collaborate through leadership, facilitation, training and incentives.

Implementation

This approach was applied through 4 workshops with partners and client at different stages of project, and through clusters in project structure (Figure 19).

Figure 19. Cluster structure in one of AMEC's projects

Benefits

Benefits of prime partnering are that a team is created; personnel and company "needs & wants" are established; project team goals are established; client is delighted; and the team approach is confirmed.

Future development

Future development will focus on whole life costing exercise, strategy for supply chain selection and developing of continuous improvement.

Contact presenter: Mr. Allan Neal
AMEC Capital Projects,
Salters Lane, Sedgefield, Stockton on Tees
Co. Durham. TS21 3EE
Tel: 01740 620161; Fax: 01740 621587
E-mail: allan.neal@amec.com;
Web site: www.amec.com

Mr. Robert Francis, Ontrack

Developing an IT Business

Challenge
The challenge was to circumvent the traditional supply chain with its hierarchical structures, which is in many cases ineffective and not cost efficient in meeting the needs of small and medium sized contractors.

Innovative approach and solutions
The company was launched in August 1999 the intention being to improve reliability in the construction sector in terms of delivery of products to small and medium contractors, hence the title Ontrack which denotes - *order now, track your order over night*. This improved reliability is intended to cut the amount of time wasted by skilled tradesmen in acquiring products essential to their services.

The service provided by Ontrack allows the small and medium sized contractor working on a site to place a order via a catalogue, (either via the web – Figure 20, or referring to a small brochure), by phone, fax or online.

Figure 20. Ontrack web site home page

Implementation

Benefits
The success of this ordering system has been ascribed to the following:
- The product catalogues have been kept *simple* in order to facilitate ordering, and not burden customers with massive amounts of technical information.
- A *clear, single system of discounting,* this being consistent across all product ranges has been established.
- A reliable courier service, targeting *'windows'* of delivery.

Future development
The intention is for Ontrack to place itself in such a position to make use available technology as well as more traditional approaches.

Contact presenter: Mr. Robert Francis
Ontrack Direct Limited
Unit 9, Alliance Business Park, Alliance Close
Attleborough Fields Industrial Estate, Nuneaton CV11 6SD
Tel: 0800 9753 007; Fax: 0800 9753 101;
E-mail: bob@ontrackdirect.co.uk
Web site: www.ontrackdirect.co.uk

Mr. Michael Sharpe, West Midlands Digital

Supporting best e-business practice

Challenge
A support community for e-business providing practical advice and support to companies in formulating and implementing e-business strategy.

Innovative approach and solutions
It is impossible to pick up a newspaper or trade journal these days without reading about e-commerce. The message is clear: e-business has arrived and every company and sector will be affected. E-business is about a lot more than just selling over the Net. It presents major opportunities to reduce costs, increase sales, enhance customer service and improve efficiency. These new ways of working enable firms in all sectors to move closer to their customers and suppliers. Effective exploitation of IT be a key factor in enabling firms in the construction industry to stay competitive. Many SMEs are reluctant to embrace online working, however, either because they see it as not relevant to their business or because of concerns over issues such as cost or security.

West Midlands Digital helps local firms realise the commercial benefits of digital business

Implementation
West Midlands Digital is a regional business club that aims to help local companies share their knowledge and experiences of the e-world. Through activities such as networking events, business visits, online information services and workshops, WM Digital allows local companies to keep up-to-date with the latest developments and explore what e-business means for them. The club was set up in 1999 with backing from Birmingham City Council and now has a wide range of public and private sector partners.

Benefits
Members benefit through:
- Sharing knowledge and experiences of e-business developments
- Networking with potential customers and service providers
- Benchmarking their e-capabilities and developing new skills and competences
- Access to grants and e-business support programmes

Future development
E-business is happening and all firms need to start to develop and implement a realistic e-business action plan.

Contact presenter: Mr. Michael Sharpe
West Midlands Digital
420 Birmingham Road, Marlbrook
Bromsgrove B61 0HL
Tel: 01527 877714; Fax: 01527 877714
E-mail: info@wmdigital.org
Web site: www.wmdigital.org

Dr. Lamine Mahdjoubi and Mr. Junli Yang, University of Wolverhampton

A Fuzzy Decision Support System for Materials Routing in Complex Construction Sites

Challenge
Reliable predication of materials routing is critical to the success of construction site management. Optimum forecasting for materials movement is an important consideration in the development of an effective project execution planning.

Innovative approach and solutions

The Virtual Construction Material Router (VCMR) proposes a novel approach to materials routing. It integrates advanced DSS techniques to determine the best route between various locations on complex construction sites. It is specifically designed to achieve the following objectives:

- Generate various scenarios for material routing;
- Allow rehearsal of various scenarios as construction progresses;
- Produce a critical path analysis in the planning of construction materials routing; and
- Select the most suitable route for materials movement.

A prototype of VCMR intelligent routing selection system has been developed. The routes are produced by the decision-making system, which uses a rules based fuzzy logic system to process the input criteria of routing. User defined criteria are selected or entered through a VRMC interface to supply the analytical queries. This includes destination of materials, types of materials, date and time, and materials storage locations.

Implementation

The GIS- fuzzy logic based decision support tool that is described is an ongoing research project. The prototype currently implemented is a functional system, specially designed to monitor materials routing in complex construction sites. The decision support system is based on site information, which considers the spatial allocation, scheduling and sequence of construction activities.

Benefits

This information is stored in a DBMS ready for visualisation and presentation. It is possible to execute the query function in AutoCAD MAP. The selected nodes could be linked and redisplayed in AutoCAD Map. This graphical representation will enable site managers and planner to identify possible conflicts of material movement and activities. It also allows the rehearsal of various scenarios.

Future development

While results presented here are preliminary, there is a guarded optimism about the potential of the system (Virtual Construction Material Router). With further developments, it is believed that this system will be able to perform the selection of criteria and simulate the final route of material movement in a visualisation package. In addition, there is an expectation that an effective querying system will ensure the automatic display of routes.

Contact person: Dr. Lamine Mahdjoubi
University of Wolverhampton
Wulfruna street, Wolverhampton WV1 1SB
Tel: 01902 321 000
E-mail: L.Mahdjoubi@wlv.ac.uk , J.Yang@wlv.ac.uk
Web site: www.wlv.ac.uk

Ms Margaret-Mary Nelson and Prof. Marjan Sarshar, University of Salford

Process improvement in facilities management

Challenge
Facilities are an organisation's second largest expense and can be responsible for as much as 15% of turnover. They are also the largest item on the balance sheet, typically accounting for 25% of all fixed assets. Despite these figures, the FM sector often fails to demonstrate its strategic importance to the core business. SPICE FM, a current research project at Salford University, is developing a structured learning framework to address this issue. The framework provides organisations with the capability to implement their vision by aligning the continuous improvement of FM services with the needs of the core business.

Innovative approach and solutions
The framework integrates two existing management tools to create a top-down, middle out approach to continuous improvement. The Balanced Scorecard is a strategic management system that combines traditional financial measures with operational and softer customer and staff issues, which are vital to growth and long-term competitiveness. The SPICE FM Process Maturity Model is a step-wise model that evaluates and continuously improves the management processes that support the implementation of an organisation's business strategy.

The SPICE FM Process Maturity Model is based on experiences from the construction and IT sectors which have both adopted similar approaches. The five-stage model provides a roadmap for moving FM organisations from a 'fire-fighting', chaotic process culture, to a culture of process focus and lean organisation. Each level comprises a set of key management processes that, when satisfied, stabilise an important part of the service delivery process. By following the steps in the model, an organisation can achieve step-wise implementation of its business strategy.

The SPICE FM framework is being tested in a series of case studies with major FM providers in both the public and private sectors. A case study commences with the development of a balanced scorecard for the organisation, highlighting its strategic objectives and appropriate supporting measures. Subsequently, the strengths and weaknesses of its key management processes are highlighted through assessment against the SPICE FM model. These findings form the basis for a process capability profile. The profile provides senior management visibility into the process infrastructure an organisation has to support the implementation of its vision. Improvement opportunities highlighted in the assessment are prioritised for action based on their contribution to the strategic objectives agreed at the commencement of the study.

Conclusion
SPICE FM is a pro-active approach to continuous improvement that provides organisations with the capability to implement their vision, by aligning the continuous improvement of FM service delivery processes with the strategic objectives of the core business. It supports organisational learning, and can be used to address various management sectors of an organisation.

Contact presenters: Ms Margaret-Mary Nelson and Prof. Marjan Sarshar
University of Salford, School of Construction and Property Management
Salford M7 9NU
Tel: 0161 295 3898; Fax: 0161 295 5011
E-mail: M.M.Nelson@salford.ac.uk
Web site: www.scpm.salford.ac.uk/spicefm/

Mr. John Goodall, FIEC

Is it not time for the British Construction Industry to start benchmarking itself against its more efficient continental European peers?

Challenge

A Europe wide benchmarking study drawn up by Bernard Williams Associates and published by the Financial Times in 1993 revealed that Britain's construction industry was at that time the most inefficient in Europe whereas Belgium's was the most productive. This is probably still true! Strange to say, that by all accounts, neither Sir Michael Latham nor Sir John Egan made any attempt in their proposals to explore this apparent anomaly and Bernard Williams' study contains no explanation for the differences.

Innovative approach and solutions

It is quite true, that Latham did realise that a system of "single point defects liability" or "project" insurance could solve many of the industries' ills (in particular through "liberating" innovation) but he believed that this could only be introduced through complex new legislation and consequently he discarded the idea.

We do not know whether Sir Michael took the trouble to cross the English Channel to France or better still to Belgium and find out how it works over there, but it seems unlikely. In Belgium there is no legislation as such concerning the application of "project insurance" but it is none-the-less in widespread use. In fact, contrary to popular belief, the Belgian construction industry is arguably the least regulated in the whole of Europe; there are not even any technical building regulations!

To summarise therefore, "project insurance" is not in itself an innovation. It has been alive and in use on the Continent for a long time. Rather this is a "clarion call" to the British construction industry and it's clients to wake up and pay attention to what has been going on across the Channel for many years!

Benefits

The essential advantage of "project insurance" is that both the design team and the contractor are insured against defects and failures under a single policy. In the event of a claim it is usually immaterial who is at fault. The insurance policy simply pays out. Through bridging the "fault line" between "design" and "construction", this system significantly reduces confrontation and litigation which has become the hallmark of the British way of working.

Consequently, contractors can submit competitive offers on the basis of "alternative technical solutions" ("variants" in the European public procurement directives) based on their own original ideas thus simplifying design whilst producing benefits for clients in terms of time and cost savings through INNOVATION!

Does this explain Sir Michael's call for a 30% reduction in costs whilst at the same time raising the industry's much maligned inadequate profit margins? The industry reacted in disbelief at the time. Is anyone listening now?

Contact presenter: Mr. John Goodall
FIEC
Avenue Louise 66 - 1050 Brussels, Belgium
Tel: +32 2 510 0952; Fax: +32 2 511 0276
E-mail: j.goodall@fiec.org
Web site: www.fiec.org

Ms Victoria Joy, Addleshaw Booth & Co

Contaminated Land: The New Regime

Challenge
Contaminated land is:
Any land which appears to the local authority in whose area it is situated to be in such a condition by reason of substances in on or under the land that:-
a. significant harm has been caused or there is a significant possibility of such harm being caused or
b. pollution of controlled waters is being or is likely to be caused.
The new statutory regime imposes clean-up liability on polluters and/or 'innocent' owners or occupiers of contaminated land and includes the first statutory definition of contaminated land. Local authorities must identify such land and those in England have until the end of June 2001 to devise a strategy for doing so.

Innovative approach and solutions
Following identification of a site, there is a mandatory minimum 3 month period for consultation with potential 'appropriate persons', at the end of which the local authority may serve remediation notices on them. These notices specify what is to be done by the polluter and/or landowner to remedy the problem and could require investigation, monitoring or immediate clean-up. They can be appealed against if considered unjust but will go on a public register. Failure to comply with a remediation notice is a criminal offence.

The 'appropriate person' is the:
- person who has caused or knowingly permitted the contaminating substances to be in, on or under the site, i.e. the original polluter – Class A
- where the original polluter cannot be found after reasonable enquiry, the current owner or occupier of the land – Class B

There are a number of exclusion rules of importance to property and corporate transactions. These can allow a polluting seller to pass on clean-up liability under this new legislation to a purchaser. Also, if appropriate persons agree in writing the basis on which they wish to divide the clean-up responsibility between them, the regulatory authorities will normally abide by such agreements. In the case of Class B owners/occupiers, licensees and tenants paying a rack rent will normally be excluded from liability.

Implementation
The effect of the new regime, which is created by Part IIA of the Environmental Protection Act 1990, is a potential clean-up liability for all historic polluters as well as owners of contaminated land.

Contact presenter: Ms Victoria Joy
Addleshaw Booth & Co
Sovereign House, Sovereign Street
Leeds LS1 1HQ
Tel: 0113 209 2000; Fax: 0113 209 2060
E-mail: vrj@addleshaw-booth.co.uk
Web site: www.addleshaw-booth.co.uk

Mr. Ray Robinson, Aon Risk Services

What Can Latent Defects Insurance Do for the Construction Industry?

Challenge

Several months ago I attended a talk here by John Goodall who indicated that if we are to achieve a truly competitive and innovative construction industry in Britain, then the industry needs to be freed from what he called its "chains and fetters" and provided with a single point of design and defect liability insurance i.e. latent defects insurance.

Innovative approach and solutions

Aon Risk Services latent defects insurance policy covers:

- Actual physical damage to the whole of the building but only caused by inherent defects that originate in the main structure.
- An inherent defect is one that exists prior to the date of practical completion but that remains undiscovered at that date and manifests itself during the period of the policy (10 to 12 years).
- The inherent defect may be in design, materials or workmanship.
- There is no need to prove negligence against a third party.

Implementation

The policy will usually be in the name of the developer but is freely assignable to new owners, lessees or financiers. Insurers are sometimes prepared to waive subrogation rights against architects, engineers, contractors and others (but not suppliers) on payment of an additional premium. Premium rates usually range from 0.50% to 1% rising to 2% for completed buildings. Fees range from 0.10% to 0.50% with an average of about 0.30%. Example: Contract price £2m: rate including TA fee: 1.00%. Total cost £20,000

Benefits

The benefits to the insured of buying the cover:

- Provides immediate funds for repairs and minimises disruption
- A phone call to a builder rather than a lawyer when damage is discovered
- Meets the insurance requirements being imposed by many financiers
- Can be a major advantage when negotiating a sale or letting.

The benefits for the construction project if the cover is purchased:

- Developers do not need to rely on project team's Professional Indemnity insurance
- The potential for confrontation is reduced
- Peace of mind for all the project team
- Less time and money is spent on arguing about contract conditions and warranties
- Innovation is encouraged
- Everyone can concentrate on getting the actual design right.

Contact presenter: Mr. Ray Robinson
Aon Risk Services
8 Devonshire Square, Cutlers Gardens
London EC2M 4PL
Tel: 020 7623 5500; direct 020 7882 0205
Fax: 020 7882 0208
E-mail: ray.robinson@ars.aon.co.uk

Mr. Henry A Odeyinka and Dr John G Lowe, Glasgow Caledonian University

The development of an expert system to manage construction cash flow and associated risks and uncertainties

Challenge
The aim of the research is to investigate the uncertainties and risk factors inhibiting an accurate forecast of construction cash flow. Based on an analysis of these factors, the research aims to develop a modular computer based expert system to forecast and manage construction cash flow.

Objectives and methodology
Objectives:
1. To identify and assess the uncertainty and risk factors involved in modelling construction cash flow;
2. To develop typologies of cash flow forecasting models based on classification categories suggested by analysis of risks and uncertainties involved in cash flow forecasting;
3. To construct an expert system model for the management of construction cash flow;
4. To develop a prototype expert system to manage construction cash flow;
5. To implement the developed system and test its forecasting accuracy.

Stage 1 (Objectives 1&2): In order to identify and assess the uncertainties and risk factors involved in modeling construction cash flow forecast, a questionnaire survey of construction contractors was carried out. Results showed that there were significant differences of contractors' opinions on major risk factors when analyses were carried out based on the groupings of size of firms, construction duration and procurement options. In order to develop typologies of construction cash flow models based on the classification categories suggested by analysis of inherent risks and uncertainties, it is proposed that a case study approach would be utilized.

Stage 2 (Objective 3): In order to construct an expert system model for the management of construction cash flow, an extensive review of existing expert systems addressing construction industry problems would be carried out.

Stage 3 (Objectives 4&5): In order to develop a prototype expert system to manage construction cash flow, an attempt would be made at the physical realization of the model constructed under objective 3. In developing the knowledge base module, it is proposed that the knowledge of the domain experts would be captured through structured and unstructured interviews. The knowledge so captured would be kept in the knowledge base module through an appropriate expert system programming language such as Flex, Prolog or C++.

Implementation
The developed prototype would be implemented to test its forecasting accuracy by imputing data for proposed construction projects. Where necessary, the system would be de-bugged and re-implemented for satisfactory performance.

Contact presenter: Mr. Henry A Odeyinka and Dr John G Lowe
Glasgow Caledonian University
Cowcaddens Road
Glasgow G4 OBA
Tel: 0141 331 3659; Fax: 0141 331 3696
E-mail: H.Odeyinka@gcal.ac.uk; J.Lowe@gcal.ac.uk
Web site: www.gcal.ac.uk

Dr. Farzad Khosrowshahi, Serenade

Refinements in Project Financial Management

Challenge

Despite the fact that the industry is well aware of the crucial role of project and corporate cash flow management, contractors and other parties have demonstrated little innovation in the way construction business is viewed as a financial entity. The industry continues to adhere to many old practices the benefits of which are not worthy of the resources that they absorb. Egan report sheds light on some of these shortcomings and recommends a more efficient approach.

Innovative approach and solutions

In order to generate a forecast of project cash flow, the proposed mathematical model utilises an extensive database. Then the expert users use their general experience to improve the forecast. Also, their specific knowledge about the project can help them to further refine the forecast. However, the resulting forecast may not be inline with the financial objectives of the organisation, therefore, ways of developing alternative cash flows are investigated. This is accomplished through negotiation with other parties (the client and sub-contractors). However, in preparation for the negotiation, the contractor uses the model to undertake a visual analysis of the impact of several variables that are influential in configuring project finance. The above is achieved at an early stage, easily, fast and cheaply, relying only on a brief description of the project.

Implementation

All the research leading to the development of the model have been encapsulated in a professional software that can be acquired and used.

Benefits

Construction has been recognised as a highly risky business, marked with a large number of bankruptcies and business failures. It is envisaged that a large proportion of these failures is associated with poor financial management. Therefore, a more proactive approach to financial management may yield higher financial rewards. Moreover, since at any give situation, the priorities of the parties (the contractor, the client and sub-contractors) may differ, the possibility of the win-win-win situation is a likely scenario. In other words, through negotiation, the parties can reach an agreement that can benefit all.

Future development

Programme is already underway to extend the above model to include corporate cash flow management. Therefore the collective impact of all present and future projects is examined visually and simultaneously. This will facilitate a better understanding about the financial role of each individual project, thus paving the way towards achieving an optimised project cash flow.

Contact presenter: Dr. Farzad Khosrowshahi
Serenade
52 Greenfield Gardens, London NW2 1HX
Tel: 020 8455 1392; Fax: 020 7815 7328
E-mail: farzad@khosrowshahi.screaming.net
Web site: www.serenade.org.uk

Mr. Mark Shelbourn, Construction IT, University of Salford

IT Self Assessment Tool

Challenge
Due to growing use of IT in businesses, senior managers need to assess the effectiveness of their company's use of IT.

Innovative approach and solutions
The Self Assessment Tool will help your business by providing a framework that allows you to:
- Assess your current use of IT
- Decide in which areas you may wish to improve your use of IT.
- Plan how best to achieve this improvement.

This is done by means of a Self-Assessment framework supported by more detailed Self-Assessment tables and a 'What To Do Next' table.

Implementation
The framework provides a table on which you can map your current levels of IT use. Once this has been done you can then map on the framework the levels of IT use you would like your organisation to achieve. More detailed Self-Assessment tables are provided to assist this process. The framework is included as a pull out for you to photocopy and use over again. The final stage is to begin to plan how to move from the present to the desired state. This is done using the 'What to do Next' table which is located alongside the Self-Assessment framework. Three main areas of IT use are considered, each of which is split into subsections. These are:
- Gaining Work – Marketing, Winning Work, and Customer Support.
- Doing Work – Briefing and Design, Planning, Procurement and Construction, Facilities Management and Maintenance.
- Supporting Work – Financial Management, Human Resources, and Business Planning.

Benefits
The Self Assessment Tool has been designed for senior managers. By using it you can make an assessment of the effectiveness of your company's use of IT. In conjunction with other Construct IT literature, How to Develop an Information Strategy, How to Implement an IT Strategy and Measuring the Benefits of IT innovation – it will help you plan the use of IT within your business. The tool is aimed at small to medium sized enterprises within the construction industry. It is a simple, yet effective, and has been designed so that companies can benefit easily through its use. The tool has been designed for repeated use to enable you to improve continuously your business practices.

There is a document entitled IT Self-Assessment Tool that can be obtained from The University of Salford that highlights the Self-Assessment Framework and various tables to enable a successful self assessment of your companies present position in the world of IT requirement.

Contact presenter: Mr. Mark Shelbourn
University of Salford
Bridgewater Building, Salford, M7 9NU
Tel: 0161 295 4793; Fax: 0161 295 5011
E-mail: m.a.shelbourn@pgr.salford.ac.uk
Web site: www.construct-it.org.uk

Mr Carl Abbott, Construct IT

How to Develop an Information Strategy Plan

Challenges
Appropriate use of Information Technology can deliver real business benefit to SMEs in the construction industry. However, if a company's use and procurement of IT systems is unplanned these benefits are unlikely to be achieved. The challenge therefore for SMEs is to introduce strategic thinking into their use and procurement of IT.

Innovative approach and solutions
Construct IT has been producing a series of *'How to Guides'* as part of the IT Construction Best Practice programme (ITCBP). Taken together the guides provide SMEs with a portfolio of guides detailing how they can obtain the maximum business benefit from their use of IT. Developing an Information Strategy is the first phase in this process.

Implementation
The guides are available free of charge to companies that register with the ITCBP. The strategy guide outlines a step by step procedure by which companies can develop an information strategy that is aligned with their business strategy.

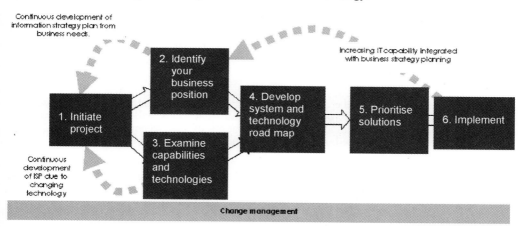

Benefits
By developing an information strategy plan SMEs can produce a framework for action, to bridge the gap from where they are to where they want to be.

Future development
Future *'How to Guides'* will detail how to achieve benefits from uses of IT such as virtual reality, website development, intranets and communications.

Contact the innovator: Mr Carl Abbott
Centre for Construction Innovation
1st Floor, 113-115 Portland Street
Manchester M1 6FB
Tel: 0161 295 5317; Fax: 0161 295 5011
E-mail: c.abbott@salford.ac.uk
Web sites: www.construct-it.salford.ac.uk
www.ccinw.com

Partners in Innovation:
Construct IT, ITCBP

Prof. Ghassan Aouad, University of Salford

Open Systems for Construction

Challenge
OSCON (open Systems for Construction) was a DETR funded project which mainly aims at developing demonstrators for applications such as architectural design, cost estimating, and planning integrated within a central database.

Innovative approach and solutions
The main contribution is the development of a set of interfaces between an object oriented database (object store) and commercial products such as AutoCAD 13, Superproject, and Netscape Navigator in order to allow the sharing of information and to allow for more user friendly interfaces. In addition to this, two utilities have been added, a Project Manager to manage communication and to monitor the progress of different phases of the construction project. A Virtual Reality (VR) interface to visualise and manipulate the design components in 3D environment.

Benefits
The main benefits associated with OSCON are as follows:

- The integration of demonstrators with an integrated database ensures that communication in the construction sector is improved resulting in better productivity.

- These demonstrators cover important areas such as architectural design, and time and cost planning.

- The popular commercial packages used in OSCON are tailored to the practitioner's need, thus savings in the form of time and cost can be expected.

- The development of the demonstrators within an established framework ensures that such a development is not done on ad-hoc basic. This will allow the construction practitioner to use the developed application in the scenario suitable for his/her business, as a result the business becomes more competitive.

- The development of models and classes ensures that a large number of practitioners and researchers can use these if further development is required. This will ensure that effectiveness is achieved.

- The development is done on a PC within a window environment. This ensures that the developed product is within the reach of most construction firms.

- The participants on the project have become aware of recent technological advances, which can help their businesses.

- The demonstrators, once adopted by the participants, will be developed into the commercial stage. This ensures that the participants will be driving the demonstrators to their own benefits.

It is concluded that OSCON has a lot of potentials, which can offer to the UK construction industry.

Contact presenter: Prof. Ghassan Aouad
University of Salford
The School of Construction and Property Management
Bridgewater Building
Salford M7 9NU
Tel: 0161 295 5176; Fax: 0161 295 5011
E-mail: g.aouad@salford.ac.uk
Web site: www.scpm.salford.ac.uk

Mr. Robert Shiret, National Design Consultancy (NDC)

Virtual Reality presentations in business: using VR to win more business

Challenge
Is it possible to use virtual reality (VR) in a variety of sectors, ranging from residential to commercial and industrial scenarios?

Innovative approach and solutions
The National Design Consultancy (NDC) is the ultimate in building engineering services, with one particular area of focus being the use of virtual reality (VR) in a variety of sectors. The range of VR applications is as follows:
- Concept or project visualisation
- Space Planning
- Architectural and industrial modelling
- Creative animation design
- Bespoke sales and marketing projects
- Communication
- Induction and training
- Security CCTV simulation
- Innovative presentations.

The medium can either be photographic stills, videos, CD-ROM's or Internet/Intranet.

Implementation
One of the key projects undertaken by the NDC was a housing development project undertaken for Mansell. New houses were being built and Mansell wanted to show their customers what the site would look like before it was built. A video was produced showing the whole site and different types of houses and amenities. A series of stills were also produced, one of which was an aerial photograph showing the site with the new development superimposed on it.

Benefits
- Tool to gain planning permission
- Brings projects to life
- Used to obtain board approval
- Enhances reputation
- Visualisation means there are time and cost savings
- Material can be used as sales and marketing tools to win work
- More effective communication
- Innovative training tools – reduces downtime
- Optimise number of security cameras

The essence is the earlier VR can be utilised in a project the more benefits the customer can achieve and therefore the more cost effective it is.

Contact presenter: Mr. Robert Shiret
National Design Consultancy (NDC)
Technology Centre, Wheatstone Road
Dorcan, Swindon SN3 4RD
Tel: 01793 494 839; Fax: 01793 480 772
E-mail: ndc.vrteam@ndc-uk.co.uk
Web site: www.ndc-uk.co.uk

Mr. Guillermo Aranda, University of Reading

Priming Novice Site Operatives Using Navigable Movies

Challenge

A high proportion of British construction activities involve work on existing buildings and structures and this demands hazard perception skills which span new build as well as refurbishment projects. Firms therefore need to ensure that their operatives have a thorough understanding of the characteristics of such working environments.

Innovative approach and solutions

Surrogate travel refers to the ability to manoeuvre through an environment without being physically present. This is achieved by capturing images in a preplanned manner generating a series of virtual nodes linked through sensitive areas so that the trainee can explore a series of settings in a random manner on the screen moving at will in a number of directions.

Implementation

Areas of investigation include ground level access/egress, above ground level access/egress, route links (e.g. crawling boards, bridges, runways and gangways), fragile roofing materials (e.g. roof lights, asbestos, metal liner panel), demarcation, guards and edge protection systems, lighting and signage systems, confined spaces, site house keeping (tidiness, obstructing barriers).

After a walk through production, the operator is then able to control the walk through sequence, with a number of possible directions at any given point (Figure 21.).

Figure 21. A segment of navigable movie

Benefits

The potential for computer-based simulations to be effective training tools have the following benefits:

- They provide a safe environment within which operatives can experience outcomes.
- They allow contractors to develop visual documentation.
- They can be effective tools to support training programs.
- They allow access to cases that would otherwise be very difficult to experience.

Contact presenter: Mr. Guillermo Aranda
University of Reading
Department of Construction Management and Engineering
Whiteknights, P.O. Box 219
Reading RG6 6AW
Fax: 0118 931 3856
E-mail: kcr97ga@rdg.ac.uk
Web site: www.rdg.ac.uk

Mr. Ian Smith, University of Northumbria

Working with multi-cultural, multi-disciplinary team

Challenge

A survey of employers of graduates from the School of the Built Environment at the University of Northumbria at Newcastle identified a range of desired skills in graduates/employees. Respondents to the survey were asked to grade the importance of key skills on a scale of 1-5 with 1 being not important and 5 being highly important. 29 Key skills were given for employers to rate with the following skills scoring rating of above 4: written communication, oral communication, problem solving, working independently, scheduling/ time management, evaluating own work performance, self control/calm under pressure, and working within a team. This indicates that the emphasis put on key skills by employers tends to be predominantly on the softer, personal non-technical skills. We see the challenge to education to be to develop learning strategies that address these key areas.

Innovative approach and solutions

We evaluated project models from within the University of Northumbria as well as other institutions and ultimately adopted the approach of involving Final Year Quantity Surveying students in establishing expectations and objectives at the start of each of the project stages and then reflecting upon the process after the event.

Implementation

The BSc (Hons) Quantity Surveying Final Year Project had to enable the development and demonstration of technical knowledge and also allow for the raising of awareness of issues relating to effective teamwork. The key to making the project unit effective was to effectively tackle the student perception of the difficulties of working in groups together with the impact that ineffective contributors could have on the assessment. This was achieved by being very honest with the students about the philosophy of the project in terms of resolving technical problems, raising awareness of team working issues, and by not attempting to ignore the problems of group projects particularly relating to assessment. The student teams were composed in such a way to allow each student to experience working with as diverse a range of ability, attitudes and cultures as possible. To be effective the teams would have to adopt strategies that established the different skills, and competencies of the members and overcome the potential problems of culture, communication and behaviour.

Benefits

The feedback received from both the students and from industry representatives has been very positive.

Future development

We believe that research should be ongoing in establishing appropriate skills profiles for employees, how these skills can be established, developed within training programmes and career progression mapped.

Contact presenter: Mr. Ian Smith
University of Northumbria at Newcastle
Ellison Building
Newcastle upon Tyne, NE1 8ST
Tel: 0191 227 3534; Fax: 0191 227 3167
E-mail: ian.smith@unn.ac.uk
Web site: online.unn.ac.uk

Mr. Matthew Finnemore, University of Salford

Structured Process Improvement for Construction Enterprises (SPICE)

Challenge

The UK construction industry does not have a recognised methodology or framework on which to base a process improvement initiative. The absence of guidelines has meant that any improvements are isolated and benefits cannot be co-ordinated or repeated. The industry is unable to systematically assess construction process, prioritise process improvements, and direct resources appropriately.

Innovative approach and solutions

SPICE is a current research project looking to address these issues by developing an evolutionary step-wise process improvement framework. The framework has been developed utilising experience from the IT sector, which has adopted a similar approach. The research has drawn specifically on the use of the Capability Maturity Model (CMM). The model is based on the maturity of an organisation's processes. Each level comprises a set of key processes that, when satisfied, stabilise an important part of the construction process. Each level lays successive foundations for the next. The model states that little value is added to the organisation by addressing issues at a higher level if all the key processes at the current level have not be satisfied. With the model, effective and continuous improvement can be achieved based on evolutionary steps. The SPICE framework can distinguish levels of increasing process capability, as shown in the SPICE Capability Maturity Model. An organisation will lie in one of the five process maturity levels, namely: Chaotic, Planned and Tracked, Well Defined, Quantitatively Controlled and Continuously Improving.

Implementation

For a Key Process to be deemed mature, it must satisfy all five of the following Process Enablers: Commitment, Ability, Activities, Evaluation, and Verification.

Benefits

Assessment of a construction organisation at a particular level will be of value to:
1- Potential clients - in assessing the suppliers' ability to deliver on time, to cost and with required quality.
2 - the organisation itself to enable it to:
- Identify and prioritise process improvement to progress to the next level. SPICE will assist in deciding which process improvement projects to select, within an EFQM framework.
- Identify improvements not to be undertaken (not feasible or high risk due to inadequate maturity).
- Compare its capability with similarly assessed organisations across the industry.
- Recognise when their organisations are ready for introduction of new technology; in order to minimise its risk.

An improvement strategy drawn up from SPICE assessment will provide well founded, low risk direction for both process assessment and improvement.

Contact person: Prof. Marjan Sarshar
University of Salford, Bridgewater Building, Salford, M7 9NU
Tel: 0161 295 5317; Fax: 0161 295 5011
Email: spice@sun.scpm.salford.ac.uk
Web site: www.scpm.salford.ac.uk/spice

Mr. Andrew Fleming, University of Salford

Process Protocol 2

Challenge
Numerous Government Reports (Banwell, Latham, Egan) have sought methods to improve the performance of the construction industry, process improvement being one such method. The Process Protocol project began in 1995 to address this issue and published The Generic Design and Construction Process Protocol in 1998.

Innovative approach and solutions
A high level process map was developed that incorporates innovative construction and manufacturing principles. The work was funded by the EPSRC, BT, Alfred McAlpine, BAA, Waterman, Capita, Advanced Visual Technology and the BRE. The Process Protocol consists of six key principles; Whole project view; Progressive Design Fixity; A consistent Process; Stakeholder/Involvement Teamwork; Co-ordination and Feedback. These principles were applied to the Process Protocol Map which can be viewed at www.processprotocol.com The Process Protocol Level 2 has developed the processes within the high level map to a higher degree of detail and developed an IT tool that supports the creation of bespoke process models.

Aims
1. To influence 'process' thinking throughout the entire construction industry, including process, people and technology issues.
2. To indicate to the industry how to embrace 'rethinking construction' principles.

Objectives
1. To develop 'sub processes' of the Process Protocol.
2. To illustrate the technology to exploit processes.
3. To illustrate an organisational framework for process improvement.
4. To illustrate implementation and change management issues related to 'process' improvement.

Benefits
The project matches the key objectives of the Innovative Manufacturing Initiative. The work helps in establishing and identifying client/stakeholder needs and improves the design and construction process by integrating technology and management. The Process Protocol and its sub process definition will address issues identified in the Construction Round Table (CRT)/ Agenda for Change. These include design techniques and processes that address customer needs, identifying and managing risk, continuous and sustained improvement in design, collaborating with suppliers and the employment of efficient construction processes.

The key driver of 'integrated processes and teams' identified in the Egan Report is also addressed. As such, a generic process will contribute to cultural change and improved communication and process management between the fragmented groups within the construction industry.

Contact presenter: Mr. Andrew Fleming
University of Salford
Centenary Building, Research Unit
Peru Street, Salford, M3 6EQ
Tel: 0161 295 6171; Fax: 0161 295 6174
Email: a.j.fleming@salford.ac.uk
Web site: www.processprotocol.com

Mr. Eric Johansen, University of Northumbria

The Government KPIs and North Tyneside's Detailed Performance Indicators

Challenge

The Government suggests a series of measurements, Key Performance Indicators (KPIs) which cover 10 areas; seven are indicators of project performance (client satisfaction with product, client satisfaction with process, defects, predictability of cost, predictability of time, actual cost, actual time), and three are measures of company performance (profitability, productivity, safety).

North Tyneside Schools Partnering Agreement has 'demonstration project' status within the government initiative Movement for Innovation, and the partnership activities will be considered against KPIs.

Innovative approach and solutions

A methodology was considered to allow development of Detailed Performance Indicators (DPI's) rooted in the Partnership Charter, but which could be related to the KPIs. The process used sought to engender *ownership* by involving as many individuals as possible in *workshops*. The objective was to clarify expectations. This clarification was used as *context* for the DPIs. The workshops contributed to the development of the structure for using DPIs to *measure* and *improve*.

Implementation

The Partnership Charter and its ethos produce expectations and outcomes which could be measured. The teams developed these and produced specific measures and proposed a structure to measure and take action to improve.

Benefits

The process enabled measurements and improvement towards achieving the KPIs and engendered ownership of the partnering process among the participants.

DETAILED PERFORMANCE INDICATORS
1. **Communication**
2. **Partnership success**
3. **Client Satisfaction**
4. **Value improvements**
5. **Health and Safety**
6. **Training improvements**
7. **Quality improvements**
8. **Manage project time effectively**
9. **Cost predictability**
10. **Time predictability**

Future development

Future research will focus on the DPI's in action and the development of performance improvements

---

Contact presenter: Mr. Eric Johansen
University of Northumbria
School of the Built Environment, Ellison Place
Newcastle upon Tyne NE1 8ST
Tel: 0191 227 4720; Fax: 0191 227 3167
E-mail: eric.johansen@unn.ac.uk
Web site: http://be.unn.ac.uk

Dr Peter Thompson, Integrated Environmental Solutions Ltd

The Evacuation Modelling Software: Simulex

Challenge
Health and safety issues in building design can be addressed through modelling of different situations and behaviour of building occupants. The challenge was to design a software package which models the evacuation of occupants from buildings.

Innovative approach and solutions
The evacuation modelling software 'Simulex' offers reliable modelling of occupant evacuation from buildings through the following features:

- Fast import of CAD-generated DXF floor plans to define each level of a building
- Automatic calculation of travel distance
- Automatic generation of escape routes
- Each person modelled individually
- Animated plan views of evacuation (Figure 22).

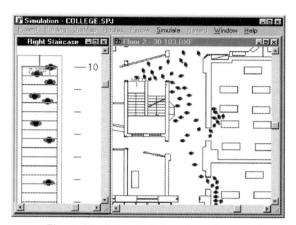

Figure 22. Animated plan view of evacuation

Implementation
The software can be used to model the evacuation of occupants from large, geometrically complex multi-storey buildings.

Benefits
'Simulex' enables building designers to accurately predict occupants' movements during an evacuation of a building, and to provide adequate escape routes and exits.

Future development
A prototype version of 'Simulex' which uses fire/smoke analyses has been developed to model the health degradation of occupants, and subsequent escape-route changes. This version will be released commercially. Also, additional options will be available to specify more exact components for both physical and behavioural occupant characteristics.

Contact presenter: Dr Peter Thompson
Integrated Environmental Solutions Ltd
141 St James Road
Glasgow G4
Tel: 0141 226 3662; Fax: 0141 226 3747
E-mail: pete@ies4d.com
Web site: www.ies4d.com

Dr. Iain Cameron, Glasgow Caledonian University

Construction 'Total Safety Management': A Benchmarking Framework

Challenge
The Health and Safety Executive 1999 Statistical Bulletin revealed that the construction industry's fatal accident incidence rate is about five times worse than the all-industry average and about three times worse than manufacturing. Historically, attempts to improve the industry 's record have focused on engineering and/or legislation. These approaches are useful but they are only part of the answer. It is now generally accepted that safety management has entered a new era which may be termed "Human Factors".

Innovative approach and solutions
The Management safety Audit, Operative Safety Inventory, and Safety Attitude Questions represent leading indicators of safety performance. Furthermore, they are amenable to measurement: attitudes (person factors) may be measured via a questionnaire (safety climate instrument); behaviour (job factors) may be measured via an inventory of site conditions (checklist of site hazards); and the situation (organisation factors) via a system audit (safety management audit).

Implementation
For the purposes of measuring what actually happens on site, it is reasonable to de-emphasise attitude studies as these can be best described as preliminary diagnostic tools. This means that 'operative behaviour' and 'management actions' collectively can be construed as approximating safety culture, which allows a 'Project Safety Culture Benchmark' to be calculated. This eclectic perspective is ideal for intra and inter organisational comparisons and in this respect is compatible with HSE's Benchmarking for Safety i.e. acting on lessons learned from internal and external best practice (= feedback). It is further possible that performance could be included within manager's annual appraisal and may even be linked to organisational rewards (promotion) to motivate additional improvements.

Benefits
A case study recently completed demonstrated an improvement in the 'Project Safety culture Benchmark' from '70 % -Safe' during the baseline phase of measurement only (months 1, 2, 3); which rose to over '90 %-Safe' during the intervention or goal setting and feedback phase of measurement plus motivation (months 4-9).

Future development
These include the extension of the technique to cover the 'Cinderella' areas of construction – occupational health. Categories of measure based on physical, chemical, and biological health hazards remain a priority e.g. Vibration White Finger, Musculoskeletal Injuries, Occupational Asthma and Respiratory Sensitisation, Allergic Dermatitis etc.

Contact presenter: Dr Iain Cameron
Glasgow Caledonian University
Cowcaddens Road
Glasgow G4 0BA
Tel: 0141 331 3244; Fax: 0141 331 3696
E-mail: I.Cameron@gcal.ac.uk
Web site: www.gcal.ac.uk

Mr. Gregory Carter[1], Dr. Simon Smith[1] and Mr. Jim Turnbull[2], University of Edinburgh[1] and Carillion plc[2]

IT Tool for Safety Risk Management

Challenge
Most drawbacks of existing risk logs stem from the fact that risk ratings are based upon individual estimations and opinions, which makes them very subjective in nature and susceptible to an individual's 'risk propensity'.

Innovative approach, solutions and implementation
The proposed IT tool uses historical data to establish the risk associated with each hazard. Consequently risk can be evaluated quantitatively rather than estimated qualitatively. The risk rating will vary depending upon the actual safety situation on-site because it is calculated from accident/incident data and updated as accidents/incidents occur.

Data from accident/incident reports and existing risk ratings were analysed. Although, at present, only a small number of accident/incident reports are available to us the results of the analysis indicates that there is a variance between the datasets. These preliminary results indicate that the existing subjective method of assessing risk does not correspond to risk determined from historical data and thus does not reflect the risk that operatives are actually exposed to on-site.

While it is true that an element of uncertainty exists in any assessment of risk, due to the fact that we are trying to predict future events, it is also true that providing a historical and factual basis for assessing risk can reduce the degree of uncertainty. Building a database of accidents will allow objective measurement of accident frequency and severity, which should lead to a more objective assessment of risk.

Overall Design Structure of the IT Tool
The IT tool will have a three-tiered structure. The first tier is the client side of the system and will be a windows-based user interface, which represents the only 'visible' part of the tool. The third tier will be a database containing the combined knowledge and experience of all personnel within the company. The second tier will consist of an application that can take a client request, interrogate the database and return it to the client side of the system. Knowledge Discovery in Databases (KDD) and Data Mining (DM) tools could be incorporated into this second tier to aid in data extraction and analysis processes.

Database management is recognised as being fundamental to the final IT tool. Data entry and access to up-to-date information represent two important operational features. A possible solution for medium to large projects, i.e. those likely to have their own servers on-site, may be to transfer the company's centralised database from the intranet to the local network drive for each individual project. Storing the database on the intranet will allow data from all projects to be sent to one single location thus solving data entry and storage problems. Downloading the database to each project's localised network reduces the effects of two common problems of accessing data on an intranet, i.e. multi-user access and bandwidth limitations. However, the database will need to be re-downloaded at regular intervals throughout the project duration to make sure that personnel always have access to up-to-date information.

Contact presenters: Mr. Gregory Carter, Dr Simon Smith
School of Civil and Environmental Engineering
University of Edinburgh, The King's Buildings, Edinburgh EH9 3JN
Tel: 0131 650 5720; Fax: 0131 650 6781
E-mail: g.carter@ed.ac.uk or simon.smith@ed.ac.uk;
Web site: www.civ.ed.ac.uk

Dr Paul Yaneske, University of Strathclyde

Greencode

Challenge

Since the UK National Health Service is a £20 billion per year business, this represents the potential for and enormous impact on the environment. A typical hospital has all the services of a small town and, as such, has the ability to make a significant impact on the environment. This alone sets it apart from the majority of industries. The fact is that some f the substances which cross the boundary of a hospital can be extremely toxic, infectious and radioactive, adding to the necessary comprehensiveness of any system developed for healthcare.

Innovative approach and solutions

Greencode is a computerised environmental management system developed by NHS in Scotland and Health and Social Services in Northern Ireland. Greencode contains a comprehensive database of UK legislation and guidance enabling a user to identify impacting legislation in a fraction of the time which would be needed to manually search through over 800 pieces of legislation and guidance.

The four countries which comprise the UK each have slightly different legal systems and as such legislation may apply in one but not in another. To avoid confusion, in the Greencode legislation database, each item is tagged with its territorial applicability. The user selects their country at the start of a review and then only legislation applicable to that country is returned in any search.

Implementation

In February 1998 Greencode assisted three hospitals, two in Scotland and one in Northern Ireland to become the first hospitals in the world to achieve certification to ISO 14001 using a generic system. Users can choose their environmental aspects from a pre-defined list and use these to link to the legislation and guidance database to gain an immediate picture of the legislation which impacts on their operations. Management system audits can be compiled on computer and analysed and presented in a range of ways. The results of these audits can then be used as a basis for prioritisation of work, monitoring and targeting.

Benefits

The Greencode software assists an organisation in two ways; it reduces the manpower necessary for the implementation of an environmental management system and contains the necessary structure such that development work is not required.

Future development

Work on the extension of the Greencode system to cover occupational health and safety has begun. It is intended that a fully integrated occupational health and safety and environmental management system will be developed.

Contact presenter: Dr Paul Yaneske
University of Strathclyde
Department of Architecture
131 Rottenrow
Glasgow G4 0NG
Tel: 0141 548 3097; Fax: 0141 548 4893
E-mail: p.p.yaneske@strath.ac.uk
Web site: www.strath.ac.uk

Mr. George Pye, Yorkshire Electricity

What is Green Electricity?

Challenge

Increasing concern about the environmental impact of the energy used in buildings has led to the development of 'green electricity' and a focus on improved energy efficiency.

Innovative approach and solutions

Renewable electricity can be generated from a wide variety of resources, such as on-shore and off-shore wind, solar power, agricultural and forestry wastes, small hydro power stations and tidal power.

Yorkshire Electricity offers its customers the opportunity to buy green electricity, which is generated from renewable resources. Currently, 75% of the company's green electricity comes from biomass resources (agricultural and forestry wastes) and 25% from on-shore wind. Due to unprecedented demand, supplies of green electricity are limited.

Benefits of green electricity

Switching to green electricity can make a positive contribution to a business's corporate environmental policy and convey a valuable message to customers about their commitment to protecting the environment. Customers can play a vital part in expanding the market for green electricity by supporting the generation of renewable energy.

Green electricity can also help organisations that have obligations to reduce carbon emissions to meet their targets.

Energy efficiency

As well as offering green electricity, Yorkshire Electricity has developed a number of sources of energy efficiency advice. A 'Focus' folder is available with practical tips on energy efficiency for large businesses as well as a leaflet containing advice aimed at small to medium-sized businesses. Advice is available via the Internet on the company's online energy efficiency directory at www.yeg.co.uk/business.

The company also offers a new energy management product called Energy Minder. This offers energy efficiency audits for larger businesses, in partnership with energy management companies that can deliver quality analysis and advice.

Contact person: Josi Mackinnon
Yorkshire Electricity
Wetherby Road, Scarcroft
Leeds LS14 3HS
Tel. 0113 289 5601
Fax 0113 289 5170
Email: josi.mackinnon@yeg.co.uk
Website: www.yeg.co.uk/business

Mr. John Mulholland, Nifes Consulting Group

Energy and Water Management at LMU

Challenge
Reduction and efficient use of energy and water on property estates contribute to their sustainability. What policies and processes should be changed to achieve this aim? How they should be implemented?

Innovative approach and solutions
NIFES Consulting Group campaigns aims to save energy and water in organisations by raising the awareness and motivation of end-users via:
- awareness raising materials,
- training,
- policy and
- strategy.

This information is then incorporated into existing corporate processes. This will aid in transforming an organisation's overall trends of energy and water consumption.

One key element of NIFES innovation is to give a quantitative score to 'Motivation and Awareness' levels of people to save energy and by plotting these scores on a graph so that potential savings can be identified.

Implementation
Each individual is plotted that if average assessment of 'Motivation and Awareness' is 50 / 50 and a campaign moves people to 80 / 80 an 8% energy saving can be made. The assessment enables organisations to predict the outcome of a campaign before it is run thus justifying the expense and reducing risk.

NIFES have designed and developed over 40 energy and environmental campaigns. From each campaign they have incorporated new knowledge in the next one so the product is continually improving. For instance, by identifying the dynamics affecting change in organisations effort can be focused in key areas.

Benefits
The product is easily measured against the costs. The amounts saved far out weigh the cost of implementing the product. A typical cost saving ratio is 1:10, i.e. £10 saved in energy costs against £1 invested in the campaign. Furthermore, the measures contribute to significant reductions of CO_2 emissions.

Future development
As each campaign is run, new data is fed into systems to refine the predictability model.

Contact presenter: Mr. John Mulholland
NIFES Consulting Group
Spinney Hill, Landmere Lane
Ruddington
Nottingham NG11 6ND
Tel: 0115 984 4944
Fax: 0115 984 4933;
E-mail: training@nifes.co.uk
Web site: www.nifes.com

Mr Dave Hampton, ABS consulting

Continuous Commissioning

An operation focussed process for delivering:
- the optimum of cost and value to occupiers
- a productive, safe and enjoyable environment to occupants

Challenge

Very few buildings work as initially intended by their design teams. As responsibility passes from client, to designer, to contractor, to occupier, to the maintenance team, there is significant opportunity for things to go wrong, for misunderstandings, and for strategy to give way to practical expediency. It is estimated that poor commissioning typically results in increased energy costs of 10 – 20%, and failure to commission can increase energy costs by 40% or more.

Commissioning (and re-commissioning) is an often over-looked professional activity that can help extract maximum benefit from a building for minimum resource input. The disciplines of good commissioning practice will have a renewed significance over the coming decades as facilities and estates managers attempt to extract more from less. Continuous commissioning is the foundation of an optimum operation strategy for buildings. It will ensure that buildings achieve best practice standards and deliver the optimum of cost and benefit to occupiers.

Innovative approach and solutions

This project introduces an innovative way to approach and market the need for better commissioning of buildings throughout their life-cycle. The aim is to develop a practical, effective and continuous technique, known as 'continuous commissioning' - a process, focussed on operation, by which a building and its services are conceived, designed, constructed, commissioned, operated, maintained and decommissioned to provide the optimum of cost and value for the occupier.

Implementation

The project brings together building owners and occupiers, designers, commissioning specialists and training providers, to ensure that the technique developed will be clear, practical and simple to use. Case studies will be produced to illustrate the approach.

Benefits

Happy people, improved productivity. Fit buildings: buildings that are fit for purpose, fit for people, fit for the planet. Sustainable businesses.

Future development

The technique will underpin the growth of the embryonic new service industry focused on delivering a total service – delivering comfortable and effective working environments to clients rather than just buildings, or just maintenance, *i.e. buildings as a service rather than as a product.* This is a topic at the centre of progressive sustainability thinking.

Contact presenter: Mr Dave Hampton
ABS consulting
6-8 Marshalsea Road, London, SE1 1HL
Tel: 020 7378 0006; Fax: 020 7378 0016
E-mail: drhampton@absconsulting.uk.com
Web site: www.absconsulting.uk.com

Innovative partners:
ABS consulting; Nationwide Building Society; Unison; Abbey National Plc; Commissioning Specialists Association; Commtech Limited; Mid Career College; CIBSE; DETR

Mr. David Taylor, Kier Group

A Corporate Environmental Policy

Challenge
Environmental and sustainability issues came to prominence following the Bruntland report "Our Common Future" (1987) and The UN Conference "Environment and Development" in Rio de Janeiro (1992). Policies and actions affecting the environment have been changing through legislative and regulatory instruments and pressures from non-governmental organisations, prompting the UK construction industry action.

Innovative approach and solutions
Kier Group is one of the UK's largest construction companies with 6,5000 employees world wide and £ 963 M turnover in 1999. The Group has developed a structure for implementation of its environmental policies (Figure 23)

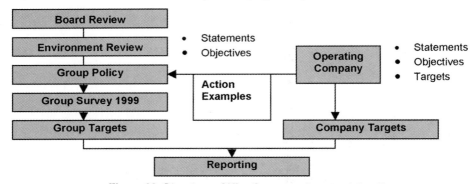

Figure 23. Structure of Kier Group environmental policy

Implementation
Examples of action/research initiatives are as follows: Partnering for Environmental Training (Kier - UH), Environmental Body Steering Group (BRE and others), Contaminated Land Risk Group (BRE, MU and others), Sustainable Energy Source Research Support, Waste to Energy Research, and many others.

Benefits
Kier Group environmental policy contributes to its competitiveness, and to the positive impact of environmental, social and economic effects of its activities.

Future development
Future development will focus on whole-life costing/ life-cycle assessment, innovation, design choice, ethics and social responsibility, efficiency, ecology, reduced use of resources, pollution prevention, and profit to enable investments.

Contact presenter: Mr. David Taylor
Kier Group Head Office
Tempsford Hall
Sandy, Bedfordshire SG19 2BD
Tel: 01767 640 111
Fax: 01767 640 002
E-mail: david.taylor@kier.co.uk
Web site: www.kier.co.uk

Dr Branka Dimitrijević, University of Strathclyde

Durability, Adaptability and Energy Conservation (DAEC) Assessment Tool

Challenge
Durability, adaptability and energy conservation of buildings are among the most important features in design of more sustainable buildings. The challenge was:
- *to develop the metrics for these qualities,*
- *to test them on the existing higher education buildings*
- *to identify conflicts between the examined qualities and*
- *to propose a strategy for their reconciliation in the form of design guidelines.*

Innovative approach and solutions
The main outcome of the research project is the DAEC Tool which assesses durability, adaptability and energy conservation of buildings. The Tool was tested on selected higher education buildings and on a new building design project for a community hospital which was designed by HBG Ltd. The Tool was developed as software which is available on a CD together with other outcomes of the research.

Implementation
It is envisaged that the DAEC Tool (available on a CD as software) will be used in the following stages of building design and building management:
- Development of design brief and identification of achievement goals.
- Building design.
- Client evaluation of building designs.
- Comparison of the whole life costs of different options for achieving the quality targets
- Assessment of the durability, adaptability and energy conservation of existing buildings.

Benefits
The DAEC Tool offers aid in defining a required building performance profile in the design brief. When the achievement goals and quality targets have been built into the DAEC tool, the design team can use the DAEC tool for the assessment of different design options. The evaluation tool can be used when a client needs to assess the projects submitted by different design teams. The Tool enables comparison between the desired quality targets and the costs for their achievement The assessment can also be used in the decision making process on maintenance, upgrading, functional improvement, changes of use and potential savings.

Future development
Future research could focus on refining the DAEC Tool through its application in building design situations for different types of buildings, and for client evaluation of different building designs in the bidding process.

Contact presenter: Dr Branka Dimitrijevic
Department of Civil Engineering, University of Strathclyde
107 Rottenrow
Glasgow G4 0NG
Tel: 0141 548 4688
Fax: 0141 553 2066
E-mail: branka.dimitrijevic@strath.ac.uk;
Web site: www.ce.strath.ac.uk/sustain

Dr. Eric Whale, JRA Aerospace & Technology Ltd

Space: A Technology Source

Challenge

Space technologies involve a variety of different disciplines and sectors within research and development and offer a vast range of technologies for exploitation.

Innovative approach and solutions

Space activities have created a diversity of technologies, which are available to non-space organisations through the European Space Agency (ESA) Technology Transfer Programme. e.g. software for modelling, analysis, requirements handling, mission control; new materials (composites – CMCs, C/SiC, high temperature alloys, polymers); insulation (knitted polymer textile resistant up to 1150°C); power (photovoltaic cells, power management systems); electronics (communications); imaging (Gamma-ray imaging, x-ray, photon-multipliers); sensors (radiation monitors, gas analysis, vibration detectors).

Implementation

An example of a technology developed for space but finding applications on Earth is SPADD – Smart Passive Damping Device.
Technology: Light weight energy dissipating device which does not alter mechanical characteristics of a structure. Turns vibrational energy into another form of energy.
Space use: Reduction of vibrations transmitted to the satellite by the launcher.
Earth use: Concrete mixer manufacture. Manufacturers were looking to decrease noise, and tried several approaches. None were successful due to conditions. Successfully applied SPADD. Currently manufacturing the first systems.

Benefits

The ESA Technology Transfer Programme has given significant benefits to Europe
- By the end of 1998 4mEuro revenue to space technology donors and 20mEuro revenues to space technology receivers.
- By the end of 2000 at least 700-800 jobs created or saved.
- Since 1990 at least two companies created.
- By end of 1998 at least 12mEuro in taxpayer return.

Access to space technologies

Spacelink Group – four core organisations in Germany, France, Italy and UK managing a network of correspondents in the ESA member states and Canada.

Space technology is being promoted through Technology Catalogue and web site www.technology-forum.com.

Contact presenter: Dr Eric Whale
JRA Aerospace & Technology
JRA House, Taylors Close
Marlow SL7 1PR
Tel: 01628 891 105
Fax: 01628 816 58;
E-mail: eric.whale@jratech.co.uk
Web sites: www.jratech.com
    www.techbank.com

Mr. John Gilbert, John Gilbert Architects
Sustainability in architectural practice

Challenge
Shettleston Housing Association is based in the East end of Glasgow and has a good track record of providing housing for low-income groups. This particular project was won in a competition set by Scottish Homes on the theme of providing sustainable housing.

Innovative approach and solutions
The housing project, completed in 1999, was designed and built with the aim of achieving sustainability in the construction of the building, by the careful selection of materials and in the energy efficiency of the houses, by providing a heating system which uses a combination of geothermal and solar energy. Geothermal heating has for too long been considered only appropriate for areas with hot springs, but with the increasing amount of disused underground coalfields, there is a plentiful supply of low-grade energy waiting to be tapped. The principals and mechanics of the system are fairly simple as the technology is not new. It is the combination of energy supply and efficiency which makes this scheme unique, yet it can be easily repeated.

Implementation
Our main aim in the Shettleston (Figure 24) proposal has been to reduce tenants running costs whilst ensuring that the construction of the houses and selection of materials takes into account the full implications of their energy lifecycle. A Focus Group of future tenants was involved in design and specification issues. Individual were offered a 'tenant's choice' option which allowed each tenant to make selections from a variety of choices within a set budget.

Figure 24. Shettleston housing

Benefits
The houses have a high level of thermal insulation, passive ventilation with reducing air infiltration. Materials were selected for their low embodied energy and reduced toxicity. Geothermal energy combined with supplementary solar panels provides inexpensive heating. Running costs for heating and hot water are estimated to be in the region of £90 a year for each tenant. The house designs allowed increased internal flexibility and loft spaces were adaptable. The houses are built to barrier free standards, creating a varied mix of house types, from family, to young persons and the elderly, along with wheelchair use. The development is car free.

Future development
The architects have a rehabilitation scheme on site in Fife which is also heated using geothermal energy from disused coalmines.

Contact presenter: Mr. John Gilbert
John Gilbert Architects
6F3 Templeton Business Centre
Glasgow G40 IDA
Tel: +44 (0) 141 551 8383; Fax: +44 (0) 141 551 9123;
E-mail: john@johngilbert.co.uk
Web site: www.johngilbert.co.uk

Douglas Taylor, Assist Architects Ltd.

New Energy Efficient Housing in Ayr

Challenge

This new build housing project at Mainholm Road in Ayr comprises 20 houses for rent, completed in September 1999. The client, West of Scotland H.A. invited us to prepare a scheme for a competition run by Scottish Homes on the theme of Energy Efficiency, which we won with a SAP rating estimated at around 114.

Innovative approach and solutions

Not only are the buildings compact in form, service runs are closely grouped, reducing heat loss through pipe runs. Hot water and heating are provided by condensing combi-boilers which are, unsurprisingly in such well insulated buildings, grossly oversized. Six of the units have an innovative heat recovery system instead. All twenty houses have low energy lighting and the conventionally heated houses use passive-stack ventilation. Materials were, where practicable selected for their "greenness". A lightweight steel frame was used which is both recyclable and indeed has itself been made largely from recycled steel. The steel frame was fabricated locally thus minimising transport costs. Another example of "green" product selection was the use or Orientated Strand Board in lieu of plywood. Locally produced high performance softwood double glazed windows were used for the same reason. External wall construction is steel frame, overclad with insulated render giving U-values of 0.18 and minimising cold bridging. Great attention was paid to air tightness with 6 properties site tested. A "prototype" unit was tested at an early stage whereby both detailing and workmanship were put to the test.

Implementation

Adaptations to each house can be readily carried out to provide a wheelchair accessible shower room (pre-plumbed for future use), hoist connection from ground to upper floor, direct connection from bathroom to bedroom. The project is currently being monitored to confirm the effectiveness of the energy efficiency measures described above.

Figure 25. New housing in Ayr

Benefits

New homes save energy and provide flexibility. All houses have ground floor toilets, large walk in stores and rooms which allow for a variety of future adaptations.

Future development

Assist Architects are developing other sustainable housing projects.

Contact presenter: Mr. Douglas Taylor
Assist Architects
100 Kerr Street
Glasgow G40 2PQ
Tel: +44 (0) 141 554 0505
Fax: +44 (0) 141 554 6112
E-mail: douglast@assistarchitects.co.uk

Dr. Hassan Al-Nageim, John Moores University, Liverpool

New Product Innovation

Challenge

Recycling of building materials is important for more sustainable construction. The challenge is to develop new processes and products for upgrading waste & low quality aggregates for use in bound & unbound road and airfield flexible pavements.

Innovative approach and solutions

A new technology and production method has been developed to address this problem by applying a coating paste to the surface of natural and presently unsuitable aggregate. The paste alters the aggregate surface properties so that the restriction on their use can be reduced.

Chemical Testing

Uncoated aggregates were compared with coated aggregates and the coating material to determine whether or not the coating could provide an improved affinity with bitumen. The test used was the Net Adsorption test developed in the USA for the SHRP. The test procedure was wholly in accordance with BS DD226: 1996 to check the deformation of bituminous road mixture under the action of road traffic.

Summary of Findings

In general the findings of this research are extremely promising. The coating paste behaves well during chemical testing, proving itself to be amenable to bonding with bitumen, and very resistant to the degrading effect of water ingress. Net adsorption, representing resistance to stripping, was improved by a factor of more than two.

The physical properties of the aggregates tested were substantially improved with the addition of coating paste The angle of internal shearing resistance and surface roughness of the aggregates were also improved considerably, thereby enhancing the strength of bituminous bi-products.

Similarly elastic stiffness and resistance to axial deformation were improved by quite significant factors with the addition of the coating paste.

Benefits

The new products are *considerably* cheaper than the primary aggregates.

Future development

Liverpool John Moores University is looking for an industrial partner to produce the product for industrial use at both a national and international levels.

Contact presenter: Dr H. Al-Nageim
School of the Built Environment
Liverpool John Moores University
Clarence Street, Liverpool L3 5UG
Tel: 0151 231 3265
Fax: 0151 709 4957
E-mail: h.k.alnageim@livjm.ac.uk
Web site: www.livjm.ac.uk

Mr. Graham Meller, ELE International

A Standardisation Trial for the Analysis of Asphalt by the Ignition Method

Challenge
The analysis of the components of an asphalt mixture have traditionally been carried using solvents to extract the binder from the aggregate. The solvents used have varied, but all are relatively toxic, harmful to the environment, expensive, and their use is generally deprecated.

Innovative approach and solutions
The concept of determining the binder content of an asphalt mixture by evaluating the loss of mass when a sample is burnt, leaving the aggregate to undergo a grading analysis, has been suggested as a means of avoiding the use of harmful solvents (other than when samples of the binder are required for subsequent analysis). Equipment for the analysis of asphalt by means of ignition (termed "Asphalt Content Tester" using the American term for bitumen) has recently been developed. An Asphalt Content Tester is an analyser which determines the asphalt content of a sample by loss on ignition whilst the asphalt sample is bathed in oxygenated air. The Asphalt Content Tester has:
- an integral balance to allow the sample to be weighed continuously throughout the ignition procedure; and
- an integral computer with software which:

- identifies the end point of ignition;
- indicates the completion of the test; and
- down-loads the results to a printer, and/or to a PC

Implementation
The method has gained relatively wide usage in the USA very quickly and is currently being standardised. As part of the exercise, a standardisation trial was carried out to determine the precision of the method. The equipment has now been introduced into the United Kingdom and a British Standards Institution *Draft for Development*, based on the American drafts, is being prepared to codify its use in the United Kingdom. To encourage the move away from the use of solvents, The Environmental Best Practice Programme commissioned TRL to run a UK standardisation trial with ELE International Limited providing the participating laboratories with the use of the ELE Asphalt Content Testers. The trial of the ignition method to estimate binder content in asphalt mixtures was required to establish the repeatability and reproducibility of the methodology. A copy of the Paper is available from graham.meller@eleint.co.uk

Conclusions
- Repeatability & reproducibility comparable with solvent method
- Precision equally applicable to loose or compacted samples
- Gaseous emissions of carbon were not significant

The BS has now been published - BS DD 250

Contact presenter: Mr. Graham Meller
ELE International Ltd
Chartmoor Road, Chartwell Business Park,
Leighton Buzzard, Beds LU7 4WG
Tel: 01525 249 200; Fax: 01525 249 249
E-mail: graham.meller@eleint.co.uk

Mr. Graham Woodall, G.D. Woodall and Company

Seeing Opportunities

Challenge
Is it possible to recognise and create a niche market for future work out of existing work with a client?

Innovative approach and solutions
Graham Woodall of G.D. Woodall and Company has through an acquired knowledge of insurance and in particular loss adjusting, created additional work for his company out of existing business with clients. The innovation lay in recognising the processes involved in insurance and loss adjusting and what skills might be required to perform in this area. The realisation was that a lot of the work done by external loss adjusters for Mr. Woodall's clients, could be adequately done by himself. The reason for this, being that he had developed a lot of the knowledge and information necessary to undertake this type of work through exposure to his clients' business.

Implementation
The only real problems faced by Mr. Woodall were trying to inform himself adequately as to the nature of the work that was required – this necessitated a lot of research into insurance principles and terminology. In addition a portion of the client base, namely other firms of loss adjusters became competitors, and were therefore 'lost' as potential clients. The innovation is not strictly transferable, as it operated within a specific context, which has subsequent to the innovation also changed quite substantially. The aspect that is transferable is the thinking behind the innovation and about using whatever opportunities that are available to transfer into another sector of the construction industry, particularly one that may not be traditional.

Benefits
In addition to increased volume of work the benefit of a wider client base is significant to any small enterprise. Further, the acquired knowledge introduced variety to the normal workload and also led to meeting people from other industries.

Future developments
As a concept, the innovation could be repeated be acquiring wider skills. Over the past few years we have seen building societies becoming banks, banks becoming insurance brokers or pension consultants, electricity companies selling gas, gas companies becoming internet service providers, supermarkets becoming all of those things as well as selling cars and the petrol to put in them so why should construction consultants remain restricted to traditional activities?

As far as the loss adjusting work is concerned, the changes in the instructing insurance market that have taken place over the past few years have made this much less attractive. Insurers now demand more and more for less fees and services demanded tend to be on a conveyor belt system rigidly controlled by QA systems with only secondary regard for efficiency and professionalism. Any future development is therefore likely to be in a different field.

Contact presenter: Mr. Graham Woodall
G.D. Woodall and Company
16a Market Street, Lichfield
Stratfordshire WS13 6LH
Tel: 01543 418 941; Fax: 01543 263 926
E-mail: gdwoodall@constructors.org

Mr. S. Christopher Smith, Smith Brothers (Tamworth) Developments Limited

Marketing through clients

Challenge

As a small property investment and development company we had observed during the 1980's that tenants were seeking shorter more flexible lease terms. It became a policy to seek to increase the retention of tenants within the portfolio as they grew or changed their business. This would reduce our marketing costs, reduce the number of vacant premises and the length of void periods between lettings.

Innovative approach and solutions

In order to be able to retain tenants it is essential to know how their business is performing and where they are taking it. By knowing this we can tailor property to their needs both physically and with the use of flexible leasing arrangements. This requires a direct contact with the tenant with a view to building a long term relationship. Trust in the landlord being able to deliver what they have said they can, is essential.

Implementation

The use of flexible leases allowed us to tailor lease length to the different requirements of different industries. Programs for the improvement of older industrial properties were put in place with the aim not at obtaining a quick increase in rent but a longer term growth. The company has also continued to undertake a program of speculative property development with units of a variety of sizes. Tenants have been kept informed of this program so that buildings can be customised to their needs, or building programs reorganised to provide space at the time best suited to the tenant. This has also demonstrated to the tenant the landlords commitment to improvement in the quality of buildings.

Benefits

The benefits of this style of property management have been reduced periods of vacancy for properties, and therefore an increase in income. The flexibility that can be offered has resulted in tenants wanting to grow within our portfolio of property. Tenants have been retained and increased in size as their business has developed. One of the biggest benefits has been the number of referrals that have resulted from existing tenants, and prospective tenants who were never accommodated. The good name of the business and the style of property management is something that other businesses value. By communicating with the tenants and building relationships, the flexibility we offer in leasing property has allowed our tenants to grow their business while we continue to grow ours.

Future development

Future development of the innovation include more flexible leases for specific areas of business and the use of new information mediums as a means of communicating with our Tenants.

Contact presenter: Mr. S. Christopher Smith
Smith Brothers (Tamworth) Developments Limited
Riverside Estate Office
Atherstone Street, Fazeley, Tamworth, Staffs B78 3RW
Tel: 01827 289 951; Fax: 01827 251 901
E-mail: SBDTAMLTD@aol.com
Web site: www.tamworthproperties.co.uk

Prof. John A. Cantwell, University of Reading

Innovation in Smaller Firms

The traditional view
Large firms are more innovative because:
- Their market power creates the profits and security needed to undertake risky projects.
- They have to rely more on non-price competition vis-à-vis other large firms.
- There are economies of scale in R&D.

The evidence
- The largest firms do tend to be the most R&D-intensive. However:
- This measure favours large firms, since they are more likely to have formalised R&D divisions, while for smaller firms innovation comes under 'design' etc.
- Industries with more giant firms tend to be the most R&D-intensive (e.g. oil, electrical)

The new view
- Within industries innovation tends to be higher in smaller firms as well as large firms - it is 'U-shaped' (Figure 26).Industries are different: smaller firms are most important as innovators in specialised machinery and instruments, but have only a weak role in building materials (or electrical equipment, chemicals, mining or defence).

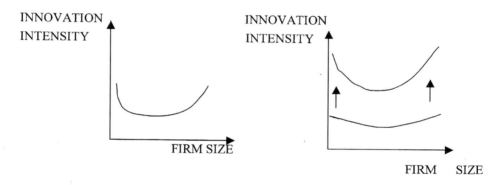

Figure 26. A U-shaped relationship of innovation intensity and firm size

Why smaller firms succeed?
- Usually not because of an 'enterprise culture' independent of and competing with large established firms
- Due to links with the skills and strategies of large firms in the same industry, on whom they depend for expertise, knowledge and markets – either as users of innovations or as collaborators (providing spin-offs).

Contact presenter: Prof. John A. Cantwell
University of Reading, Department of Economics
PO Box 218, Whiteknights, Reading RG6 6AA
Tel: 0118 987 5123; Fax: 0118 975 0236
E-mail: J.A.Cantwell@rdg.ac.uk
Web site: www.rdg.ac.uk

Ms. Emma Buxbaum, Construction Industry Council

Raising the competence of SMEs in the Construction Industry

Challenge
- To enable SMEs to make full use of Occupational Standards and NVQ/SVQs by being <u>aware of their uses and benefits</u> and the available support for them.
- To generate <u>demonstrations of good practice</u> in the use of Occupational Standards and NVQ/SVQs and disseminate the good practice.
- To develop <u>good mechanisms for supporting SMEs</u> in the use of Occupational Standards and NVQ/SVQs and disseminate them to intermediary organisations.

Innovative approach, solutions and implementation

Four target audiences:
- Contractors
- Specialist Sub-contractors
- Consultants
- Local Authorities

Support mechanism in the form of:
- Telephone help line
- On-line resource centre
- Presentation team

Consultancy SMEs:
- Development of a Training Needs Analysis
- Establishment of Best Practice Clubs

Contracting SMEs:
- Support for on-site assessment
- Develop contractor/sub-contractor network
- SMEs working together sharing mentor/assessor services
- Evidence matrices for qualifications

Specialist Sub-contractors:
- Promote the Roofing Industry Alliance Hallmarking Scheme
- Identify and train a network of mentor/assessors to support its implementation
- Establishment of Assessment Centre for specialist sub-contractors

Local Authority SMEs:
- Optimisation of networking activities
- Sharing of assessment services
- Development of evidence matrices for qualifications
- Production of toolkit to meet 'Best Value' requirements

Benefits

Benefits to:
- Process
- People
- Expenditure/Income

Future development

The way forward is detailed in a number or practical toolkits and guidance materials for organisations working with SMEs (and larger companies) and the companies themselves. These are available from the Construction Industry Council.

Contact person: Mr David Cracknell Construction Industry Council Standards Committee
The Building Centre, 26 Store Street
London WC1E 7BT
Tel: 020 7637 8692; Fax: 020 580 6140
E-mail: mail@cicsc.org.uk
Web site: www.cicsc.org.uk

Colin Pearson, BSRIA

Feedback for better building design and construction

Challenge
The Probe series of building reviews published in Building Services Journal, has now reported on 22 buildings between one and five years after occupation. The problems that occur all too often can be summarised in terms of essential features that are often absent or poorly implemented.

A Vail Williams survey in 1990 found 65% of users dissatisfied with the heating, ventilation and air-conditioning. 45% of claims and litigation are due to errors in design concepts. Defects cost the construction industry a billion pounds a year. Occupiers are frustrated with little scope for feedback. High profile reports from Latham and Egan have also stressed the need for feedback.

Innovative approach and solutions
What feedback is in use today? A BSRIA survey found 37 forms of feedback in use among building design consultants including: Pre-Briefing Appraisal, Feasibility Review, Value Engineering, Client Satisfaction Survey, Commissionability Review and Post Occupancy Evaluation. But these were islands of feedback; nobody was using feedback across the board in all aspects of their work and at all stages in projects. The research showed that whilst the benefits of feedback are recognised, there are many barriers to the effective uptake of feedback within the industry.

Implementation
The project went on to develop, with the project partners, ways of overcoming many of the procedural barriers related to the mechanism of carrying out feedback. Methods included setting up formal feedback systems, holding training sessions, making feedback part of quality system, agreeing mutual sharing of feedback at start of each project, reviewing the company feedback system and adopting a no-blame culture. However the barriers related to motivational factors are more intractable and require a significant cultural change within the industry.

Benefits
Project partners found a reduction in defects, improved solutions, a more cost effective process, improved client liaison, enhanced building performance and higher productivity.

Future development
BSRIA has progressed to developing a mechanism for technical design quality assurance and working closely with the facilities management industry to generate feedback on building operation. BSRIA also runs in-house training for companies wishing to implement feedback.

Contact presenter: Mr. Colin Pearson BSRIA Old Bracknell lane West, Bracknell Berkshire, RG12 7AH, UK Tel: +1344 426511; Fax: +1344 487575 E-mail: colinp@bsria.co.uk Web site: www.bsria.co.uk	**Innovative partners:** Usable Buildings MACE, CBX, CBS Paul Banyard and Associates AMEC Design and Construction Post Office Property Holdings N G Bailey & Co Ltd Metropolitan Police

Mr. Andrew White, University of Warwick

The Sources and Enabling Factors of Innovation

The Challenge
- Organisations are prone to staying with their (yesterdays) innovations for too long.
- Innovations are subject to limited performance levels and revenue streams.
- The greater the percentage of revenue drawn from new products the more financially successful an organisation is likely to be. **Knowledge and ideas**, not **land, labour and capital**, are increasingly becoming the means by which wealth is created and measured

Innovative approach and solutions

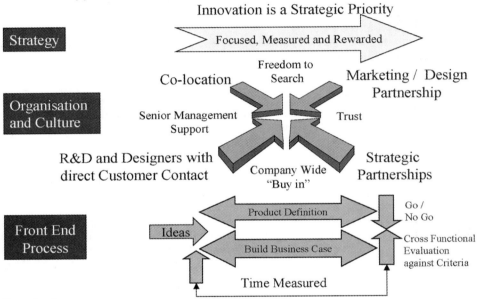

Conclusions
- Innovation is a critical factor in business success
- Ideas are critical inputs to the innovation process
- The sources from which innovative ideas emerge are often accessed in a serendipitous nature
- Factors have been identified that enable ideas to transformed into innovative concepts
    o Strategy
    o Process(s)
    o Organisation
    o Culture

Contact presenter: Mr. Andrew White
Rm 325 International Manufacturing Centre
University of Warwick
Coventry CV4 7AL
Tel: 024 7657 2346; Fax: 024 7657 2123
E-mail: a.d.white@warwick.ac.uk
Web site: www.wmg.org.uk